THE NORTH PACIFIC CRETACEOUS TRIGONIID GENUS *YAADIA*

BY

LOUELLA R. SAUL

UNIVERSITY OF CALIFORNIA PUBLICATIONS IN GEOLOGICAL SCIENCES
Volume 119

UNIVERSITY OF CALIFORNIA PRESS

THE NORTH PACIFIC
CRETACEOUS TRIGONIID GENUS *YAADIA*

THE NORTH PACIFIC
CRETACEOUS TRIGONIID
GENUS *YAADIA*

BY

LOUELLA R. SAUL

UNIVERSITY OF CALIFORNIA PRESS
BERKELEY • LOS ANGELES • LONDON

UNIVERSITY OF CALIFORNIA PUBLICATIONS IN GEOLOGICAL SCIENCES

Editorial Board: D. I. Axelrod, W. B. N. Berry, R. L. Hay, M. A. Murphy,
J. W. Schopf, M. O. Woodburne, Chairman

Volume 119
Approved for publication June 24, 1977
Issued June 30, 1978

University of California Press
Berkeley and Los Angeles
California

University of California Press, Ltd.
London, England

ISBN: 0-520-09582-0
Library of Congress Catalog Card Number: 77-84990

CONTENTS

THE NORTH PACIFIC
CRETACEOUS TRIGONIID GENUS *YAADIA*

by

LOUELLA R. SAUL

Abstract

The lineage of trigoniid species usually referred to in faunal lists as *Trigonia leana* Gabb belongs to the North Pacific genus *Yaadia* Crickmay, 1930. West Coast deposits have yielded ten species of the following ages: *Y. lewisagassizi* Crickmay, Valanginian (?); *Y. jonesi* n. sp., Hauterivian; *Y. whiteavesi* (Packard), Albian; *Y. leana* (Gabb), Cenomanian-Early Turonian; *Y. californiana* (Packard), Late Turonian; *Y. pinea* n. sp., Coniacian; *Y. branti* n. sp., Santonian; *Y. tryoniana* (Gabb), early Campanian; *Y. robusta* n. sp., Late Campanian; *Y. hemphilli* (Anderson), Maestrichtian. They can now be used to indicate the age of the beds in which they occur. Both the reduction and rounding of the anterior angulation and the increasing curvature of the dorsal margin through time—morphological changes that help make species discrimination possible—can be interpreted as adaptive to a change of living position. Shells of Early Cretaceous *Yaadia* spp. have a high anterior angulation which probably served as a stabilizer, preventing the current from disorienting the shell. By Turonian time the anterior end became rounded for easier burrowing, the somewhat more buried shells being less in need of a down-current brace. The dorsal portion of the shell developed a concave curvature in compensation for this increased burial of the ventral margin so that the incurrent-excurrent area remained above and parallel to the substrate.

The *Trigonia leana* group has been referred to *Steinmanella* but West Coast specimens of Hauterivian and Albian age clearly show that *T. leana* auct. are *Yaadia*. *Yeharella* spp. which are very similar to *T. leana* auct. are also probably *Yaadia*. *Yaadia* appears to be derived from Mid Jurassic *Scaphogonia*, but *Steinmanella* does not; any common ancestry for *Yaadia*, *Steinmanella*, and *Quadratotrigonia* is probably pre-Mid Jurassic. In the Cretaceous, four groups of knobby trigoniids have been found. Each one occurs mainly in a separate geographic area; *Yaadia*, in the North Pacific; *Litschkovitrigonia*, in the eastern Tethys; *Quadratotrigonia*, in the western Tethys; and *Steinmanella* in the southern ocean. Knobby trigoniids disappeared from the geological record first in the south, second the western Tethys, then the eastern Tethys, and lastly from the North Pacific. The morphologic similarities of these large, knobby Cretaceous trigoniids probably result from convergent or parallel adaptations to similar shallow water, sandy bottom habitats.

Introduction

This paper continues the study of the fauna from the type Chico Formation. The remarkable continuous record of the autochthonous North Pacific Cretaceous fauna makes recognition of species difficult, and in each lineage the same specific name has generally been applied to specimens from various stages. Correct identification of the knobby trigoniids from Chico Creek, thus required revision of the group through the Cretaceous. When the knobby trigoniids are distinguished as to species and genera, their record supports some theories of speciation and extinction and counters others.

These highly ornamented trigoniids, *Trigonia leana* Gabb auct., do not make an ideal group with which to work. The occurrence of the fossils is localized, even within the medium- to coarse-grained sandstone to which they are nearly restricted. The knobby valve surface seldom breaks free from the matrix, and no method of preparation was discovered that cleaned the specimen without removing some shell. The difficulty that other paleontologists have experienced in distinguishing species in the *T. leana* group can be taken as evidence of their probable close relationship, and they are inferred to show phyletic rather than allopatric speciation through the Cretaceous along the West Coast of North America. Each of the ten species delimited in this study

1

has a limited time range and can now be used to determine the age of beds where it may be found. Recognition of some of the evolutionary trends of this group led to assigning the species to *Yaadia* Crickmay, 1930b.

Inclusion of later Cretaceous West Coast species in *Yaadia* adds to the definition of the genus and suggests some generic reassignments for various species of exotic knobby trigoniids. Recorded occurrences of knobby trigoniids assigned to the four groups *Yaadia, Steinmanella, Quadratotrigonia,* and *Litschkovitrigonia* define for each a paleogeographic territory. These four groups do not simultaneously cease to occur in the geologic record, and the order of their apparent extinctions may coincide with some possible major current changes brought about by continental drift. Such slow changes can, however, hardly be a direct cause of extinction. The wide distribution of the knobby trigoniids should have favored survival as they would not be everywhere subjected to adverse physical conditions. Biotic factors such as the life habits of the knobby trigoniids, the impingement of other animals, etc., were probably more important than continental drift in bringing an end to their lineages.

Most specimens of West Coast *Yaadia* collected over the past century have been studied. Poor preservation of holotypes and missing or misleading type locality designations have caused confusion. The few specimens previously available and the considerable morphological variation made definitive specific characteristics difficult to recognize. Ten species are at this time distinguished. Of these *Y. lewisagassizi* and *Y. tryoniana* are as yet known from inadequate material. Specimens of *Y. californiana* better than the holotype were recently found, but the species is still represented by too few specimens, and more specimens are needed to provide the basis for easier distinction between *Y. californiana* and *Y. robusta.* The most easily recognized of the ten species are *Y. jonesi, Y. whiteavesi, Y. branti,* and *Y. hemphilli.*

The name most often applied to West Coast knobby trigoniids has been *Trigonia leana* Gabb, but eleven species names are available: two—*tryoniana* (1864) and *leana* (1877)—were provided by Gabb; three—*whiteavesi, californiana,* and *fitchi*—by Packard (1921); one—*lewisagassizi*—by Crickmay (1930b); and five—*perrinsmithi, wheelerensis, colusaensis, hemphilli,* and *branneri*—by Anderson (1958). All available described and/or figured type specimens were borrowed. The holotype of *T. californiana* Packard is missing from the University of Oregon collections, but the California Academy of Sciences has a plaster cast. Canada Geological Survey provided excellent plaster casts of the holotype and paratype of *T. whiteavesi* Packard, Whiteaves' hypotype of *T. tryoniana* Gabb from Northwest Bay, British Columbia, and a latex pull of the holotype of *Yaadia lewisagassizi* Crickmay.

The distinctive shape and complex sculpture of trigoniids have given rise to an abundance of morphological terms. Those terms that have seemed most appropriate for discussing the knobby trigoniids are indicated in text figure 1. Although "area" is more commonly used (e.g., Cox *in* Moore, 1969, N471), *corcelet* has also been employed (e.g., Newell and Boyd, 1975, p. 66) for that portion of the shell dorsal to the marginal angulation. The *marginal angulation* (Stoyanow, 1949, p. 67), which separates the corcelet from the flank, is far from marginal and therefore somewhat confusing; but it is better suited to knobby trigoniid morphology than marginal carina (Cox *in* Moore, 1969, N106) or posterior ridge or carina (Newell and Boyd, 1975, p. 66) which suggest a more abrupt feature than is present on these forms. The corcelet is

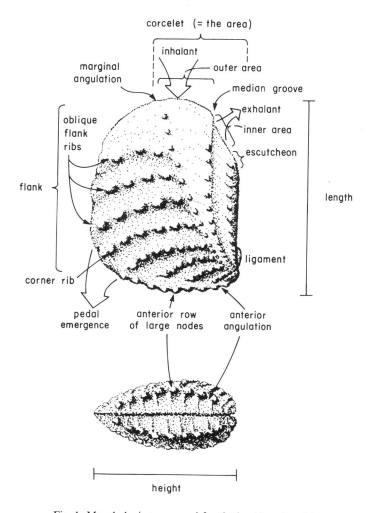

corcelet (= the area)

inhalant

marginal
angulation

outer area

median groove

exhalant

inner area

oblique
flank
ribs

escutcheon

flank

length

corner rib

ligament

pedal
emergence

anterior row
of large nodes

anterior
angulation

height

Fig. 1. Morphologic terms used for the knobby trigoniids.

divided into a dorsal wedge or *inner area* and a ventral wedge or *outer area* by a *median groove*. The corcelet marks the respiratory margins of the animal: the outer area valve margins indicate the position of the inhalant current; the inner area valve margins that of the exhalant current of the animal. The height on specimens of *Yaadia* spp. is measured perpendicular to the ligament and the length parallel to the ligament.

As a result of their shell structure and preservation in well-cemented sandstone, the trigoniids from the West Coast have an unfortunate tendency to peel, leaving an outside layer of shell in the matrix and some shell on an internal cast. In order to retain the sculpture, it was necessary to grind the matrix from around the nodes using small diamond wheels. Because the shell is much softer than the matrix, most specimens so prepared are somewhat pock-marked by the grinder even though the preparation was done under binocular magnification. The fine sculpture about the beaks usually flakes away as the matrix is removed and is preserved on only a few West Coast specimens. This sculpture is considered of importance in reconstructing phylogenies by such Japanese trigoniid experts as Kobayashi, Nakano, and Tashiro.

Specimens from one locality were used whenever possible to indicate the amount of variability to be expected in shape and sculpture of a species; but for most of these species it was necessary to use specimens from several localities of similar strati-graphic position.

Loans of types and other specimens from the following persons and institutions made this study possible. The abbreviations are those used in the text to indicate ownership of the specimens.

CAS, California Academy of Sciences; Peter U. Rodda

CGS, Canada Geological Survey; T. E. Bolton, T. P. Poulton

CIT, California Institute of Technology (specimens presently at UCLA)

HSU, California State University, Humboldt; Frank Kilmer

LSJU, Stanford University; A. M. Keen

UCB, UCBMP, University of California, Berkeley and Museum of Paleontology;
    J. W. Durham, J. H. Peck

UCR, University of California, Riverside; J. D. Mount, M. A. Murphy

UO, University of Oregon; L. R. Kittleman

USGS, United States Geological Survey; D. L. Jones

USNM, United States National Museum; E. G. Kauffman

Stratigraphic information on areas with which I am superficially acquainted and areas which I was unable to visit was generously supplied by W. P. Popenoe, D. L. Jones, P. U. Rodda, P. D. Ward, and J. A. Jeletzky. W. P. Popenoe and John Alderson have donated for my use all of the knobby trigoniids they could find. Drafting of the more complex and expertly inked figures was by Vicki Doyle. Ram Alkaly made the thin sections mentioned in the text. Takeo Susuki took several excellent photographs which are acknowledged in the plate descriptions. The manu-script benefited from the careful reading and thoughtful criticism of R. B. Saul, W. P. Popenoe, G. B. Cleveland, M. A. Murphy, and D. L. Jones.

Species Groups of Knobby Trigoniids

Early Cretaceous specimens of *Yaadia* are most easily distinguished from other knobby trigoniids (see text fig. 2). Early Cretaceous *Yaadia* spp. have the distinctive flattened anterior end, bordered by a strong row of nodes clearly separated from the oblique flank ribs along the anterior angulation. Early Cretaceous specimens of *Quadratotrigonia* are relatively easily recognized by their sculpture. The regularly enlarging and spaced nodes of the flank give a neat appearance even on those forms with indistinct ribs. The corcelet has fine ribbing or pustular noding. *Yaadia* and *Steinmanella* have flanks sculptured by irregularly sized and shaped nodes which produce an untidy effect despite the nodes being clearly set on oblique ribs. The corcelet of *Steinmanella* is coarsely, concentrically wrinkle ribbed. Several *Quadratotrigonia* spp. have differentiation of an anterior zone of nodes, but the anterior profile is not flattened. *Litschkovitrigonia* has flank ornament similar to *Quadratotrigonia*, but *Litschkovitrigonia* is more elongate; its corcelet lacks the pustular ornament of *Quadratotrigonia*, and its marginal angulation, median groove, and escutcheonal angulation are much less ornamented.

Because of the repetition of similar facies in western North Pacific deposits it has been possible to assemble a series of *Yaadia* spp. spanning the Cretaceous. Species of this series from the later Cretaceous are not easily differentiated on morphology from *Steinmanella*, and without this series to work from, various authors (e.g., Kobayashi and Amano, 1955; Nakano, 1960 and 1968) have included these forms of distinct phylogeny in *Steinmanella*. Cox (1952, p. 57) suppressed *Steinmanella, Transitrigonia*, and *Quadratotrigonia* as synonyms of *Yaadia* thereby arriving at the correct generic assignment for the West Coast species, but also confounding lineages that can be distinguished over a long period of geologic time.

Beginning in the late Albian, the anterior row of large nodes so distinctive of Early Cretaceous *Yaadia* was gradually reduced in size, but not eliminated before Late Cretaceous time. If this distinctive row of nodes was not acquired appreciably faster than it was eliminated, the Late Jurassic ancestor of Early Cretaceous *Yaadia* had a strong row of nodes along the anterior angulation. Ancestors of Late Jurassic *Steinmanella* must have lacked a differentiated row of anterior nodes, and ancestors of Late Jurassic *Quadratotrigonia* were somewhat irregularly ribbed on the anterior end.

Among Jurassic forms *Scaphogonia argo* and *S. charlottensis* seem likely antecedents to *Yaadia*, but several Jurassic *Myophorella* are irregularly ribbed and noded along the anterior margin, and an anterior set of discrepant nodes was not a rare development in Middle Jurassic trigoniids. In "*Trigonia*" *kutchensis* Kitchin (1903, p. 84, pl. 8, fig. 7-9) of Callovian age (Charce group, lower beds) from Keera Hill, India, the sculpture of the anterior is more strongly differentiated than usual for *Myophorella* but less so than in *Scaphogonia* or *Scaphotrigonia*. *Scaphotrigonia navis* (Lamarck) of Aalenian age from northeastern France and southwestern Germany and "*Trigonia*" *naviformis* Hyatt, from the Bicknell Sandstone of Callovian age (Imlay, 1952, p. 976), Mt. Jura, California, have exceptionally broad, flat anterior ends, bordered by a row of very large, very clearly set off nodes. They resemble early *Yaadia*

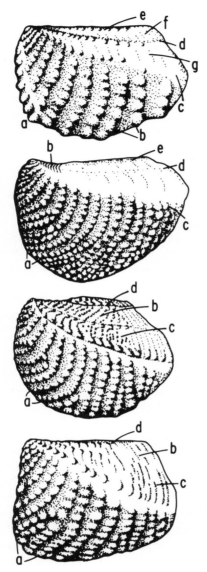

Yaadia
Truncated, angulate anterior end *(a)*, becoming rounded in Late Cretaceous species. Flank sculpture of large rude nodes on oblique ribs *(b)*. Marginal angulation *(c)*, median groove *(d)*, and escutcheon *(e)* bordered by nodes; inner *(f)* and outer *(g)* areas relatively smooth except for bordering nodes.

Litschkovitrigonia
Flank sculptured by large to medium nodes on distinct to indistinct oblique ribs *(a)*, sometimes disposed to form V's. Corcelet with fine oblique ribbing near beak *(b)* becoming smooth posteriorly. Marginal angulation *(c)*, median groove *(d)*, and escutcheon *(e)* indistinctly marked.

Quadratotrigonia
Short relative to height. Flank sculpture of medium to small nodes on distinct to indistinct oblique ribs *(a)*. Corcelet with pustular noding on inner *(b)* and outer *(c)* areas especially near beak. Escutcheon *(d)* broad and distinctly marked by angulation.

Steinmanella
Elongate relative to height. Flank sculpture of large rude nodes on oblique ribs *(a)*. Corcelet strongly wrinkle ribbed parallel to growth lines on inner *(b)* and outer *(c)* areas. Escutcheon *(d)* distinctly marked by angulation (see also pl. 11, figs. 7-8).

Fig. 2. Summary of characteristics
distinguishing four knobby trigoniid groups.

in having the row of large anterior nodes set off from the flank ribbing by a broad interspace, but their flank ribbing of clearly sharply raised, slightly nodose ribs, has greater similarity to *Pterotrigonia* ribbing rather than to the low, broad very noded ribs of *Yaadia*. "*Myophorella (Promyophorella)*" *montanaensis* (Meek), Bajocian to Early Callovian, western interior United States, is similar to *Scaphogonia* and *Yaadia* but differs from *Scaphogonia argo* Crickmay mainly in lacking the row of nodes along the median groove (Imlay, 1964, p. C32). *T.* cf. *T. spinulosa* Young and Bird and *T. undulata* Fromherz from Big Bend area, Shasta Co., California (Sanborn,

1960, pl. 2), are not illustrated showing clearly the sculpture of the anterior, but the figures suggest *"T." dawsoni* (Whiteaves), and they are probably not *Scaphogonia*. *Yaadia* cannot be directly related to all of the Jurassic forms with a distinct row of anterior nodes, and it may not be related to any yet described. *Scaphogonia argo* Crickmay, Bajocian, Ashcroft, B.C., and *S. charlottensis* (Packard), Callovian, upper part of Yakoun Formation (McLearn, 1949, p. 12), Queen Charlotte Islands, B.C., are similar to *Yaadia* and have the advantage of an eastern North Pacific provenance.

If *Yaadia* derives from *Scaphogonia*, any close connection to *Steinmanella* must be pre-mid Jurassic. Saveliev (1958, fig. 8) derives *Yaadia, Steinmanella, Quadratotrigonia*, and *Litschkovitrigonia* from *Myophorella* at about the Jurassic-Cretaceous boundary, but the Kimmeridgian species *"Trigonia" vyschetzkii* Cragin is probably a *Steinmanella*, and *"T." dumblei* Stoyanow from the same beds is a *Quadratotrigonia* (Nakano, 1960, p. 265-266; 1968, p. 34). An even earlier species whose sculpture resembles that of *Quadratotrigonia* is *"Haidaia" billhookensis* Crickmay from the Callovian Mysterious Creek Formation of British Columbia. Nakano (1968) derives Quadratotrigoniinae, which includes *Quadratotrigonia, Litschkovitrigonia* and *Korobkovitrigonia*, from the Vaugoniinae especially *Orthotrigonia*, but he derives *Steinmanella* from *Myophorella* in Myophorellinae. Early Jurassic noded Vaugoniinae of Japan are replaced by Late Jurassic Myophorellinae (Kobayashi and Tamura, 1955, p. 89) and both the stratigraphic succession and sculpture suggest phylogenetic relationship of some species in these subfamilies. The present divisions seem in part to define stages of evolution rather than phylogenetic relationship. Probably Vaugoniinae (at least in part), Myophorelliinae, Quadratotrigoniinae, and Pseudoquadratotrigoniinae should be included in Myophorelliinae as should *Quadratotrigonia, Steinmanella*, and *Yaadia* which must already have diverged into separate stocks before the end of the Jurassic. Both *Yaadia* and *Quadratotrigonia* have similar posterior lateral hinge teeth which resemble those of Early Jurassic *Vaugonia*. These unimpressive teeth suggest relationship, but they may be another instance of the parallel evolution so rampant in the trigonaceans (Newell and Boyd, 1975).

Among knobby trigoniids there has been parallel evolution leading to a straighter marginal angulation, large size, and broad, large-noded, oblique flank ribs. Bajocian *Scaphogonia argo* Crickmay and Callovian *S. charlottensis* (Packard) have a more curved marginal angulation than *Yaadia*; but the Japanese "Oxfordian or younger" (Kobayashi, *et al.*, 1959, p. 283; Nakano, 1960, pl. 28) *"Myophorella (Myophorella)" dekaiboda* Kobayashi and Tamura (1955, p. 95, pl. 6, fig. 6a-b, 7-9) appears to have had less curve to its marginal angulation, tending thus toward a more blocky *Yaadia*-like shape. The Middle Jurassic forms have an arcuate marginal angulation and concave dorsal valve margin and an almost alate projection of the dorsal valve margin immediately posterior to the beak. In *Yaadia* the ligament is short, close to the beak and the line of its attachment is along the relatively straight dorsal valve margin; the escutcheon is flat to convex and its plane makes an angle of 60° to 80° to the sagittal plane. The escutcheon surfaces of *Scaphogonia argo* and *S. charlottensis* slope concavely away from the valve margins making a very acute angle with the sagittal plane. In some other Jurassic Myophorellinae, such as specimens of *Clavotrigonia perlata lexoviensis* Chavan of Rauracian age from Calvados, France, this angle is 90°. The evolution of the nearly straight dorsal margin with inflated escutcheon was thus not concurrent in the lineages developing it. In *Y. jonesi* this angle is smaller than in later *Yaadia* spp. The angle is, generally, slightly smaller for *Yaadia* spp. than for most *Quadratotrigonia*.

The three other series of knobby trigoniids discussed in the literature all have shorter geologic ranges than does *Yaadia;* Weaver's (1931) series of South American *Steinmanella* spp. runs from Neocomian to Aptian; Saveliev's (1958) series of *Quadratotrigonia* to *Litschkovitrigonia* to *Korobovitrigonia* lasts from Neocomian to Turonian; and the Japanese series of *Yaadia* spp. discussed by various authors as *Steinmanella (Yeharella)* spp. (e.g., Nakano, 1960) is known from Cenomanian to Maestrichtian. From Texas and Arizona are *Quadratotrigonia* spp. of Late Jurassic through Early Cretaceous age (see table 1) described by Stoyanow (1949).

These southwestern species of *Quadratotrigonia* are confusing and confused. Saveliev (1958, p. 179) refers *"T." guildi* Stoyanow [=*"T." resoluta* Stoyanow and *"T." saavedra* Stoyanow], *"T." taffi* Cragin, and *mearnsi* Stoyanow to *Quadratotrigonia*

TABLE 1

Stage	*Quadratotrigonia*	*Steinmanella*
ALBIAN Early	*Q. mearnsi* (Stoyanow) [? = *Q. gordoni* (Whitney)]	
APTIAN Clansayan Gargasian (late)	*Q. guildi* (Stoyanow) [= *Q. resoluta* (Stoyanow); *Q. saavedra* (Stoyanow)]	
(early) Bedoulian		
NEOCOMIAN		
KIMMERIDGIAN	*Q. dumblei* (Stoyanow)	*S. vyschetzkii* (Cragin) [= *S. maloneana* (Stoy.); *S. maloneana* var. (Stoy.); *Trigonia* sp. Stoyanow]

Stoyanow's (1949, p. 73) erroneous selection as holotype of *Steinmanella vyschetzkii* (Cragin, 1893) of a specimen, USNM cat. no. 28967, collected in 1895 subsequent to the original description of the species causes no taxonomic confusion as the specimen selected is of the same species as Cragin's original lot. Those specimens referred to *"T." maloneana* by Stoyanow are less crushed and slightly better preserved specimens of *Steinmanella vyschetzkii* (Cragin).

The holotype of *Q. guildi* has slightly larger nodes on more distinct ribs than the type specimens of *Q. resoluta* and *saavedra*, but the differences between *Q. guildi, resoluta,* and *saavedra* are probably intraspecific variation, and the latter two based on fragmental specimens should be synonomized with *Q. guildi* (Stoyanow). A plaster cast of the holotype of *Q. taffi* (Cragin) from Bluff Mesa, El Paso Co., Texas, has nodes of similar size to *Q. resoluta* and *saavedra*, but the flank ribs are closer spaced. The stratigraphic position of *Q. taffi* is not clear. Stanton (*in* Cragin, 1905, p. 29) records it from the Quitman Formation, Quitman Mts., El Paso Co., Texas, associated with *Exogyra quitmanensis* and below beds containing *Orbitolina texana*. Stoyanow (1949, p. 37, 79) found *Q. mearnsi* in the upper Quitman Formation of Mayfield Canyon, Quitman Mts. below the *Orbitolina* zone, and Stanton's *Q. taffi* is probably *Q. mearnsi*. The top of Bluff Mesa is capped by limestone containing *Exogyra quitmanensis* and *Orbitolina texana* (Baker, 1927, p. 25). If the holotype of *Q. taffi* is from the top of Bluff Mesa, it could be younger than *Q. mearnsi* and its differences from *Q. resoluta* and *saavedra* would be of specific importance rather than a further intraspecific variation.

(Transitrigonia), using *Transitrigonia* Dietrich, 1933a, rather than *Steinmanella* Crickmay, 1930b, because of the prior *Steinmannella* Welter, 1911 (Crickmay, 1962, p. 11, has provided the unnecessary substitute *Steinmanaea* for *Steinmanella*). These have the fine ribbed to pustular ornamentation on the corcelet and the equant outline of *Quadratotrigonia* and are not *Steinmanella* [= *Transitrigonia*]. "*Trigonia*" *gordoni* Whitney from the Albian Glen Rose Formation was described as being without tubercles on the corcelet, but the illustrated beakless fragment (Whitney, 1952, p. 702, pl. 87, fig. 2) has most of the corcelet missing, and Nakano (1960, p. 266) has referred it to *Steinmanella*. It resembles *Quadratotrigonia mearnsi* (Stoyanow), however, in the size of the flank nodes which are larger than those of *Q. taffi* (Cragin) to which Whitney compared the species. Furthermore, the ornament of the corcelet of *Q. mearnsi* fades toward the posterior valve margin and a similarly fragmental specimen of *Q. mearnsi* would show no more sculpture of the corcelet than does the illustration of *Q. gordoni* which is here referred to *Q. mearnsi* (Stoyanow).

Among the fossils from the Alisitos Formation, Baja California, which were left undescribed by the too early demise of E. C. Allison, are two pulls from rock casts of *Quadratotrigonia*. One, a large fragment showing flank sculpture, is from UCB loc. B5676—a locality for which Allison never recorded the data in the UCBMP locality file; the other, a beak fragment, is without locality number. The presence of *Quadratotrigonia* is mentioned in prebatholithic Alisitos Formation, Santo Tomas valley, Field Trip Stop 3 (Allison, *et al.,* 1970, p. 132). The nodes on the flank fragment are slightly larger than those on specimens of *Q. guildi* (Stoyanow), but smaller than those on specimens of *Q. mearnsi* (Stoyanow). The beak fragment resembles *Q. guildi*. The flank fragment appears to have been from a larger specimen than any of the Arizona specimens of *Q. guildi* in the UCLA collections, and these two fragments are probably both *Q. guildi* ?=*Q. taffi* (Cragin). If they are the specimens mentioned in the field trip log, they present further evidence of the late Aptian age of the Lower Member of the Alisitos Formation. Some of the "*Trigonia*" listed from UCB loc. A8330 (Allison, 1955, p. 403, fig. 2) are "*T.*" *weaveri* Stoyanow ="*T.*" *lerchi* (Hill), a species which Stoyanow found in beds of Late Aptian age. A latex pull from another Baja California specimen which suggests *Quadratotrigonia guildi* (Stoyanow) is figured by Perrilliat-Montoya (1968, p. 15, pl. 6, fig. 7). The specimen is presumably from the vicinity of San Quintin-El Rosario, but its stated locality AOH-32, "Rosario Formation, collected by Amado Osaria Hernández," is not plotted on plate 1 nor listed with the other fossil localities, page 7. The occurrence of such a form in the very Late Cretaceous age Rosario Formation seems unlikely; its preservation also appears unusual for the Rosario Formation and more like that of the San Fernando Formation of Perrialliat-Montoya [?=Alisitos Formation of Allison (1955)]; and it is probably of Early Cretaceous age.

Distribution

Despite the inevitable lacunae in the paleontological record of near-shore bivalves, distributions of four groups—*Yaadia, Steinmanella, Quadratotrigonia,* and *Litschkovitrigonia-Korobkovitrigonia*—are plotted in text figures 3-5 for Late Jurassic-Early Cretaceous, Middle Cretaceous, and Late Cretaceous. The occurrences of

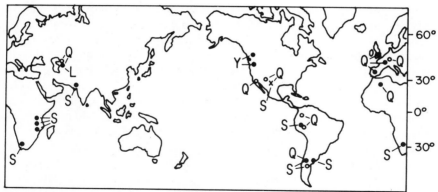

Fig. 3. Latest Jurassic and Early Cretaceous

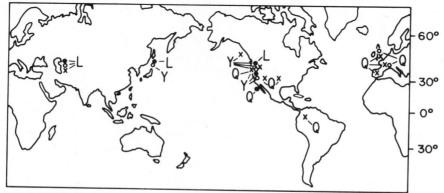

Fig. 4. Middle Cretaceous

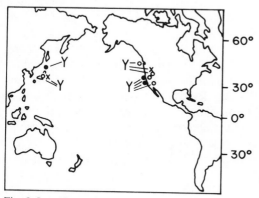

Fig. 5. Late Cretaceous

Figs. 3-5. Known occurrences of knobby trigoniids: S = *Steinmanella.* Y = *Yaadia,* Q = *Quadratotrigonia,* L = *Litschkovitrigonia-Korobkovitrigonia.* Distributional data from Coates (1974), Gillet (1965) Kitchin (1903; 1913), Kobayashi and Amano (1955), Nakano (1960; 1968), Saveliev (1958), Stoyanow (1949), Weaver (1931), and this paper. Fig. 3. Latest Jurassic and Early Cretaceous: x = Late Jurassic, • = Neocomian, o = Aptian. Fig. 4. Middle Cretaceous: x = Albian, o = Cenomanian, • = Turonian. Fig. 5. Late Cretaceous: x = Coniacian-Santonian, o = Campanian, • = Maestrichtian.

knobby trigoniids probably record only a shadow of their actual distribution. Plotted on maps of the continents as they are presently positioned, *Steinmanella,* is a predominantly southern group of the Early Cretaceous. *Quadratotrigonia* and *Litschko-vitrigonia-Korobkovitrigonia* are Tethyan groups and were most wide spread in the Middle Cretaceous. *Yaadia* is confined to the North Pacific; but within this relatively restricted geographic domain it flourished and apparently had the longest geologic range of the knobby trigoniids—Neocomian to Early Maestrichtian. It was still well represented in the Late Cretaceous when the areal distribution of the knobby trigoniids had diminished (text fig. 5).

On text figure 3 the following species are plotted as *Steinmanella:*

Kimmeridgian
  Trigonia vyschetzkii Cragin, 1893 [=*T. maloneana* Stoyanow, 1949], Texas.
Neocomian
  Lyrodon herzogii Goldfuss, 1834, South Africa (Kitchin, 1913, p. 102; Gillet, 1965, pl. 1); *T. mamillata* Kitchin, 1903, India; *T. holubi* Kitchin, 1913, South Africa; *T. steinmani* Philippi, 1899, Argentina and Chile; *T. haupti* Lambert, 1944, Argentina; *T. hennigi* Lange, 1914, East Africa; *T. neuquensis* Berkhardt, 1903, Argentina (Valanginian); *T. transitoria* Steinmann, 1881, South America; *T. t. quintucoensis* Weaver, 1931, Argentina (late Mid Valanginian); *T. t. curacoensis,* Weaver, 1931 (Late Valanginian).
Aptian
  T. t. vacaensis Weaver, 1931, Argentina (Late Hauterivian-Early Aptian).

These species fit the diagnosis of *Steinmanella* of Kobayashi and Amano (1955, p. 199); but those species generally placed in the subgenus *Yeharella* are included in *Yaadia.* This removes from the reported distribution of *Steinmanella* (Cox *in* Moore, 1969, p. N487) the Japanese and most of the western North American occurrences and leaves *Steinmanella* a mainly southern hemisphere group. Plotting the occurrences on a map of the inferred Neocomian earth (text fig. 6) increases the southern aspect as India is here in a more southern position. The duration of the genus is also thus changed from throughout the Cretaceous to Late Jurassic-Aptian. It was the earliest of the knobby trigoniid groups to disappear.

Plotted as *Quadratotrigonia* on text figures 3 and 4 are:

Kimmeridgian
  Trigonia dumblei Stoyanow, 1949, Texas.
Neocomian
  T. rudis Parkinson, 1811, Britain and Europe; *Myophorella (Myophorella) invittulina* Saveliev, 1958, Caspian region (Early Valanginian); *T. loewinson-lessingi* Renngarten, 1926, Caspian region (Early Valanginian); *T. erycina* Philippi, 1899, Argentina (Valanginian); *T. nodosa* Sowerby, 1829, Europe and eastern Asia; *T. mangyschlakensis* Luppov, 1932, Caspian region; *Quadratotrigonia (Leptotrigonia) balchanensis* Saveliev, 1958, Caspian region (Late Barremian); *T. subhondeana* Gillet, 1924, North Africa and Spain (Barremian).
Aptian
  T. nodosa Sowerby, 1829, Europe and eastern Asia; *Q. (L.) balchanensis* Saveliev, 1958, Caspian region; *Q. (L.) craveciae* Saveliev, 1958, Caspian region; *T. hon-*

daana Lea, 1840, northern South America, Spain, and Algeria; *T. spectabilis* Sowerby, 1822, Britain; *T. guildi* Stoyanow, 1949 [*=resoluta, saavedra*], Arizona, Texas and ?Baja California ?=*taffi* Cragin, 1893, Texas (Late Aptian); *T. daedalea* Parkinson, 1811, Europe.

Albian

T. spectabilis Sowerby, 1822, Britain; *T. daedalea* Parkinson, 1811, Europe; *T. mearnsi* Stoyanow, 1949 [*=gordoni* Whitney], Arizona and Texas; *T. verneuilli* Vilanova, 1863, Spain; *Quadratotrigonia* sp. A, northern California.

Cenomanian

T. daedalea Parkinson, 1811, Europe; *T. elisae* Briart and Cornet, 1863, Europe; *T. deslongchampsii* Munier-Chalmas, 1865, France; *T. padernensis* Petter-Receveur, 1955, France; *T. quadrata* Agassiz, 1840, Europe.

Turonian

Quadratotrigonia sp. B, Cedros Island, Baja California.

The species here plotted as *Quadratotrigonia* belong to several lineages; both *Leptotrigonia* and *Mediterraneotrigonia* are included. Their inclusion adds to the relative abundance of *Quadratotrigonia* spp. as compared to *Steinmanella* and *Yaadia* but does not change the distribution of *Quadratotrigonia*. Plotted on a map of the inferred Neocomian earth (text fig. 6), *Quadratotrigonia* essentially straddles the equator. Ocean current directions suggested by Gordon (1973) are in general accord with the migration suggested for *Mediterraneotrigonia* (Gillet, 1965; Nakano, 1974, p. 80) which includes the species *subhondeana, hondaana,* and *verneuilli.* Greater species diversity is usual in groups of equatorial distribution (Stehli, *et al.,* 1967, p. 457), but the disentangling of other *Quadratotrigonia* lineages may give information on ocean currents and land positions of the Early to Middle Cretaceous.

Two new species of *Quadratotrigonia* have been found among the West Coast specimens of knobby trigoniids. *Quadratotrigonia* has not previously been recognized here, but neither new species is represented by sufficiently well-preserved material for description. *Quadratotrigonia* n. sp. A is represented by two fragmental specimens from the Early Albian *Brewericeras hulenense* zone from north of Clear Creek, Shasta Co., California. One well-preserved but broken specimen from Cedros Island, Baja California, Mexico, is referred to as *Quadratotrigonia* n. sp. B. It has been assigned a Turonian age on the basis of its occurrence with a *Pyrazus* n. sp. which at nearby localities occurs with *Trigonarca californica* Packard and *"Turritella" robusta* Gabb. If it is both *Quadratotrigonia* and Turonian, it is remarkable for being from the West Coast and the last of its kind. Although Nakano (1968, p. 27, 36, 41) states that the genus flourished into the Turonian of Europe and vanished in the Senonian, he lists no Turonian occurrences.

Quadratotrigonia is thus present in the Late Jurassic and continues into the Turonian, but the greatest number of species has been reported from the Aptian, and its widest geographical expansion may have been in the Albian. Despite its wide distribution and numerous species it was the second group of knobby trigoniids to disappear from the record.

On text figures 3 and 4 the following species are plotted as *Litschkovitrigonia-Korobkovitrigonia*:

Neocomian

Litschkovitrigonia tenuituberculata Saveliev, 1958, Caspian region (Early Valan-
ginian); *T. litschkovi* Mordvilko, 1953, Caspian region (Early Hauterivian); *T.
multituberculata* Litschkov, 1912, Caspian region (Early Hauterivian); *T. ovata*
Litschkov, 1912, Caspian region (Early Hauterivian); *T. minor* Litschkov, 1912,
Caspian region (Early Hauterivian); *L. media* Saveliev, 1958, Caspian region
(Early Hauterivian); *T. inguschensis* Renngarten, 1926, Caucasas (Late Barremian).

Albian

Korobkovitrigonia korobkovi Saveliev, 1958, Caspian region (Late Albian); *K.
solida* Saveliev, 1958, Caspian region (Late Albian); *K. subamudariensis* Saveliev, 1958, Caspian region (Late Albian).

Cenomanian

T. romanovski Archangelski, 1916, Caspian region; *Steinmanella (Yeharella) jimboi* Kobayashi and Amano, 1955, Japan; *T. ferganensis* Archangelski, 1916, Caspian region.

Turonian

S. (Y.) jimboi Kobayashi and Amano, 1955, Japan; *T. ferganensis* Archangelski,
1916, Caspian region; *T. fitchi* Packard, 1921, So. Oregon-No. California; *T.
amudariensis* Archangelski, 1916, Caspian region.

Saveliev's (1958) derivation of *Korobkovitrigonia* from *Litschkovitrigonia* is the
basis for plotting the two genera as one lineage on text figures 3-4 and 6-7. This group
occurs mainly in the Caspian region, and it can therefore be considered to be a
predominantly Tethyan group as is *Quadratotrigonia.* "*Trigonia*" *fitchi* Packard is
obviously not a *Yaadia* and resembles figures of species assigned to *Litschkovitrigonia*
and *Korobkovitrigonia.* "*Steinmanella (Yeharella)*" *jimboi* Kobayashi and Amano
(1955, p. 204, pl. 12, fig. 4) is apparently based on a rock cast. As figured it is
remarkably similar to *L. fitchi* (Packard) of similar age. If these two are indeed
Litschkovitrigonia, the distribution of the group is increased to include Japan and the
West Coast of North America. Various problems of correlation make it impossible to
indicate that the apperance of *Litschkovitrigonia* was certainly earlier on the west
than on the east side of the Pacific; present data suggest rough contemporaneity of
appearance.

All known occurrences of *Litschkovitrigonia* are in the northern hemisphere; its
distribution is thus more northern than that of *Quadratotrigonia.* Like *Quadratotrigonia* it appears to have had several contemporaneous species and thus to have had
greater species diversity than *Steinmanella* or *Yaadia.* Both *Litschkovitrigonia* and
Quadratotrigonia occur in western Asia in the Early Cretaceous, but *Quadratotrigonia* occurred predominantly in the western Tethys *Litschkovitrigonia* remained
more Asiatic, and its appearance in California and Oregon indicates an eastward
expansion; whereas those of *Quadratotrigonia* on the West Coast denote western
distributional bounds. *Litschkovitrigonia* is known from the Late Turonian, and it
probably survived slightly longer than its more Atlantic, Tethyan cohort.

Plotted on text figures 3-5 as *Yaadia* are:

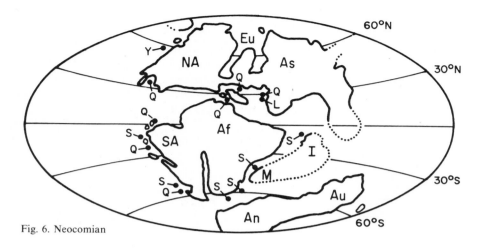

Fig. 6. Neocomian

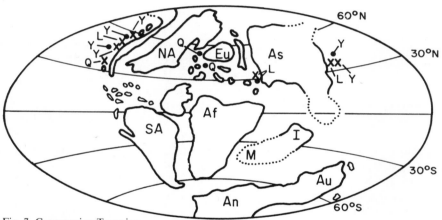

Fig. 7. Cenomanian-Turonian

Figs. 6-7. Inferred position of land in Cretaceous (modified after Gordon, 1973) and distribution of knobby trigoniids. NA = North American, Eu = Europe, As = Asia, SA = South America, Af = Africa, M = Madagascar, I = India, An = Antartica, Au = Australia, Y = *Yaadia*, S = *Steinmanella*, Q = *Quadratotrigonia*, and L = *Litschkovitrigonia*. Fig. 6. Inferred position of lands in Neocomian. Neocomian trigoniid distribution from fig. 3. Fig. 7. Inferred position of land in Cenomanian-Turonian. Cenomanian and Turonian trigoniid distribution from fig. 4, Cenomanian indicated by •, Turonian by x. In this generalized form, distributions of knobby trigoniids are equally "reasonable" with continental drift (figs. 6-7) and without continental drift (figs. 3-5).

Neocomian

Yaadia lewisagassizi Crickmay, 1930b, British Columbia (Valanginian); *Y. jonesi* Saul, Oregon (Hauterivian); *Y.* cf. *Y. lewisagassizi* Crickmay, Jeletsky *in* Coates, 1974, British Columbia (Barremian).

Albian

Trigonia whiteavesi Packard, 1921, Pacific Coast North America.

Cenomanian

T. leana Gabb, 1877, Pacific Coast North America; *T. ainuana* Yabe and Nagao, 1925, Japan.

Turonian

T. leana Gabb, 1877, Pacific Coast North America (Early Turonian); *T. ainuana* Yabe and Nagao, 1925, Japan; *Steinmanella (Yeharella) lymani* Kobayashi and Amano, 1955, Japan; *T. californiana* Packard, 1921, Pacific Coast North America (Late Turonian).

Coniacian

Y. pinea Saul, California; *T. kimurai* Tokunaga and Shimizu, 1926, Japan.

Santonian

T. kimurai Tokunaga and Shimizu, 1926, Japan; *Y. branti* Saul, California; *S. (Yeharella) japonica obsoleta* Kobayashi and Amano, 1955, Japan.

Campanian

T. kimurai Tokunaga and Shimizu, 1926, Japan (Early Campanian); *S. (Y.) j. obsoleta* Kobayashi and Amano, 1955, Japan (Early Campanian); *T. tryoniana* Gabb, 1864, Pacific Coast North America (Early Campanian); *T. japonica* Yehara, 1923, Japan; *S. (Y.) kimurai sanukiensis* Nakano, 1958, Japan (mid Campanian); *Y. robusta* Saul, California (Late Campanian).

Maestrichtian

T. hemphilli Anderson, 1958, California; *T. deckeina* Kubota, 1952, Japan.

The assignment of Japanese species to *Yaadia* is based upon descriptions and figures. Only one specimen, a plaster cast of *Yaadia kimurai* (Tokunaga and Shimizu), was available for study. This cast is of a poorly preserved specimen which lacked characterizing portions of shell. Most of the figures likewise indicate similarity to *Yaadia* but are of poorly preserved specimens or are not posed so as to display features most characteristic of *Yaadia*. Kobayashi and Amano (1955, p. 200) define *Yeharella* as lacking the transverse costellae and carina on the corcelet of *Steinmanella,* features which align *Yeharella* with *Yaadia*. None of the descriptions or figures of the Japanese species indicates the presence of large nodes along the anterior angulation. However, *Y. ainuana* (Yabe and Nagao) of Cenomanian-Turonian age is the earliest known Japanese species, and West Coast *Yaadia* spp. of similar and younger age do not have an obvious row of nodes on the anterior angulation.

Cretaceous oceanic circulation models (Luyendyk, Forsyth, and Phillips, 1972; Gordon, 1973) postulate two gyres in the North Pacific similar to those of today. It may be that the relatively limited but stable areal distribution of *Yaadia* was related to these stable current patterns. Compared to the other three knobby trigoniid groups and taking into account the length of time *Yaadia* inhabited the North Pacific, it has relatively few species. Its durability is not directly traceable to abundant speciation. The fossil record of the knobby trigoniids suggests apparent extinctions first in the south with the disappearance of *Steinmanella;* next in the western Tethys with the loss of the *Quadratotrigonia* lineages; then in the eastern Tethys as the *Litschkovitrigonia* lineages vanish; and finally, before the end of the Maestrichtian, in the North Pacific with the last of the *Yaadia* spp.

In the Early Cretaceous the knobby trigoniid form was cosmopolitan and suggests a common adaptation to a similar ecological niche. Each of the knobby trigoniid groups was relatively wide ranging but consisted of endemic species. Truly cosmopolitan species are rare especially among the benthos; organisms of more nearly cosmopolitan distribution are generally pelagic, hence the enthusiasm of stratigraphers for ammonites and foraminifera. It has been hypothesized that stocks having more restricted distribution are more likely to become extinct than cosmopolitan stocks (see Bretsky, 1973), but the benthonic habit is not necessarily equated with imminent extinction nor are pelagic animals granted perpetual survival. The earliest knobby trigoniid to disappear from the record is *Steinmanella;* its distribution—extending from India around the southern coasts of Africa and South America to Arizona (text figs. 3 and 6)—does not seem less extensive than that of *Yaadia* which confines itself to the North Pacific (text figs. 3-7).

The gradual disappearance of the knobby trigoniids may indicate the phasing out of a particular Mesozoic bivalve life style. Elements of this life style were doubtless shared by other trigoniids. The fossil record indicates a decline of trigoniid genera through the Cretaceous. On the West Coast four times as many trigoniid lineages are present in the Early Cretaceous as in the Late Cretaceous; and Nakano (1960, p. 267, table 19) indicates similar declines of trigoniid genera in Japan and the North American Gulf Coast. Ranges given in the Treatise (Cox *in* Moore, 1969, p. N478-488) suggest that trigoniid genera similarly faded away in Europe. Plotting of trigoniid generic diversity on diagrams such as Newell and Boyd, 1975, p. 58, figure 1, suggests an apparently simultaneous disaster at the end of the Cretaceous not supported by the fossil record. The gradual demise of the trigoniids is contrary to Bretsky's (1973, p. 2089) statement that extinctions have tended to occur simultaneously in all fossilized groups. More data should indicate how typical of extinction patterns is the fading of the trigoniids.

Extinction of the knobby trigoniids occurs first where postulated continental drift would cause the greatest disruption of previously existing ocean currents and last where the ocean currents remained most stable (Gordon, 1973). Valentine and Moores (1972, p. 174) have argued that continental movements must have had profound effects upon environmental conditions and thus upon biota. The climate of the Cretaceous seems to have been temperate with latitudinally broad molluscan provinces. At their coldest the Cretaceous climates were equivalent to Mid-temperate or Cool-temperate of today (Kauffman, 1975, p. 181). The slow redistribution of ocean currents brought about by continental drift would not have caused abrupt or drastic temperature changes, nor is it likely that these changes would have suddenly altered the productivity regimen which Valentine (1971, p. 259) considers a more fundamental regulator of species diversity than temperature. In the extinction of the knobby trigoniids, their own life habits may have been of greatest importance.

Ecology and Life Habits

Several geographic occurrences of *Yaadia* suggest nearshore habitat. *Yaadia* occurs near Horsetown, Shasta Co., California, in Early Albian deposits of very shallow water aspect lying upon Shasta Bally quartz diorite to which oysters adhere (Rodda, 1959, p. 61). On Bellinger Hill, Jackson Co., Oregon, the Cenomanian sandstone beds provide the most prolific *Yaadia* localities. These sandstones are in part cross-bedded and commonly contain worn fragments of oysters; also present are very large (one measured 7.6 cm. in diameter, and 40+ cm. in length) horizontal burrows. These burrows have a nearly round cross-section, are well defined by a rind-like margin, and are roughly parallel to the bedding. Similar traces have been interpreted as being dwelling burrows of crustaceans and suggestive of near shore deposition (Frey and Howard, 1970, p. 151). In the Redding area, Shasta Co., California, *Yaadia* are common at only one locality, the easternmost fossil locality of Coniacian age (UCLA loc. 6304), which is near the easternmost edge of Cretaceous outcrop. Along Chico Creek, Butte Co., California, *Yaadia* is not present in the lowermost beds which appear to have been deposited rapidly, suggesting conditions too unstable for a clam that required the growing time indicated by a full-sized *Yaadia*. *Yaadia* occur in the medium-grained sandstone beds of Santonian age overlying the basal conglomerate along with a fauna suggestive of water depths of from 7 to 40 meters (Saul and Popenoe, 1962, p. 323) and are absent from the finer-grained more basinward deposits.

   Throughout the Cretaceous *Yaadia* spp. are found in medium- to coarse-grained, poorly sorted sandstone: *Y. lewisagassizi* of Valanginian age is from a metamorphosed graywacke; *Y. hemphilli* of Maestrichtian age is from medium- to coarse-grained arkosic sandstone (Dailey and Popenoe, 1966, p. 4). Some specimens, notably *Y. jonesi,* Hauterivian, Days Creek Formation, have been found in fine-grained sandstone, but *Yaadia* spp. are more commonly associated with pebbly sandstone as in the lower part of the Chico Formation on Chico Creek, Butte Co., California, and CAS loc. 29580 in the lower Panoche Formation on San Luis Ranch, Pacheco Pass, Merced Co., California. Although sometimes found with *Yaadia, Pterotrigonia* spp. are more often from finer-grained sandstone than are *Yaadia* spp., and they have been found at more localities than *Yaadia*. This coarse-grained substrate association of *Yaadia* spp. suggests a nearer-shore, shallower-water habitat for *Yaadia*. So also does the less frequent occurrence of *Yaadia* (although specimens of *Yaadia* are locally abundant), as nearer shore deposits are less likely to be preserved than more basinward deposits. Nakano (1970) found *Pterotrigonia* in living position with the sagittal plane vertical, but no *Yaadia* has been found in living position. *Yaadia* specimens, valves together, usually have the sagittal plane parallel to the bedding plane of the sediment as might be expected for epi- to semi-infaunal bivalves of the inner sublittoral zone.

   Of the once large trigoniid group only *Neotrigonia* lives today. Its morphology, where similar to that of *Yaadia,* suggests similar life habits for *Yaadia* but, where different, indicates probable differences in life style. *Neotrigonia* has long been notorious for its leaping foot which enabled it to leap over the gunwale of Samuel Stutchbury's boat in Sidney Harbor (e.g., Fleming, 1964, p. 198). It is fairly active, moving about on the sea bottom (Allan, 1959, p. 276), and its unusually large and active foot makes possible rapid reburial (McAlester, 1965). The burial rate achieved with this foot is unusual for so plump and highly sculptured a bivalve (Tevesz, 1975, p. 325). For all its speed, it is not a deep burrower. *N. gemma* is usually buried to its valve margins; but larger specimens of *N. margaritacea,* a larger species, have epibionts attached to the posterior third of the valves, suggesting that that portion of the valves is regularly exposed (Tevesz, 1975, p. 325). The foot of *Neotrigonia* emerges near the anteroventral corner (Tevesz, 1975, p. 324), and this corner, roughly parallel to and at greatest normal distance from the hinge axis (Stanley, 1970, p. 470), is opposite the inhalant region and down (text fig. 1). In living position the corcelet margin is uppermost and parallel to the substrate. The mantle margins are unfused and *Neotrigonia* achieves separation of incurrent-excurrent flow by apposition of pallial ridges (Gould and Jones, 1974) which lie along the internal surface of each mantle lobe. The interior of each valve has a faint ridge near the margin indicating the separation between the inhalant-exhalant areas and the position of the pallial ridges. *Neotrigonia* produces large white eggs (Tevesz, 1975, p. 325), has filibranch gills, and has a vestigial byssus as an adult. The byssus probably is functional in juveniles, but the improbability that adult Mesozoic trigoniids maintained their position by byssal attachment has been clearly set forth by Gould (1969).

   Neotrigonia and *Yaadia* have similar well-developed trigonian grade schizodont dentition, similar adductor and pedal muscle scars, and their pallial lines are similarly entire, but distant from the valve margins. By analogy with *Neotrigonia* the in- and excurrent areas can readily be recognized on fossil trigoniids (Gould and Jones, 1974); the margin of the inner area was the excurrent zone of the animal (text fig. 1). The internal median ridge which is very low and short in *Neotrigonia* is longer and higher in *Yaadia*. These strong internal median ridges (pl. 1, fig. 6; pl. 3, fig. 6; pl. 7, fig. 3) assisted in the formation of functional siphons, but the strength of these internal medial ridges suggests that *Yaadia* and other fossil trigoniids probably did not have as well-developed pallial ridges as does *Neotrigonia*. The presence of the internal median ridges is a strong indication that the mantle margins of the fossil trigoniids were as free as those of *Neotrigonia*.

   Placement of the muscle scars of *Yaadia* and *Neotrigonia* suggests for *Yaadia* a similarly large L-shaped foot and perhaps equal burrowing ability. As large specimens of *Yaaida* spp. are at least twice as large as *N. margaritacea,* which generally has the posterior end above the substrate, living adult *Yaadia* were probably only semi-infaunal. Nakano (1970, p. 20) proposed that the shell of young *Nipponitrigonia* was more completely buried than that of adult *Nipponitrigonia,* and *Yaadia* may have behaved similarly. Small clams can work their way into the substrate more easily and with proportionately less energy expenditure than larger ones (Stanley, 1970, p. 54). Specimens of *Yaadia* spp. were surveyed for epibionts which might have found a semi-infaunal bivalve a place of abode. A specimen of *Y. hemphilli* (UCLA cat. no. 38639) from the Jalama Formation had an oyster attached near the posterior end of the corcelet and a specimen of *Y. branti* (UCLA cat. no. 38618) has a similarly placed bryozoan colony. Some epibionts were undoubtedly removed in cleaning the *Yaadia* shells so that the species could be identified. A number of *Yaadia* shells from various horizons are riddled with holes, possibly sponge borings. Such borings are often best developed in the posterior region, but some shells were very well drilled all over. The oyster, bryozoans, and borers do not clearly, of themselves, indicate a semi-infaunal position for the infested *Yaadia* as the oyster, bryozoan, and excavators of the borings could have moved in on the *Yaadia* shells after death of the clams if the shells lay unburied on the sea floor.

   The blunt anterior of Early Cretaceous *Yaadia,* made effectively broader by the ruff of strong nodes along the anterior angulation, seems a perverse development for a burrowing clam. It resists being pushed into sand, and the flattened anterior acts as a brake even though the shell is rocked back and forth. Late Cretaceous *Yaadia* spp. have a more rounded anterior and are more readily rocked into sand (text fig. 8). Profiles of Early Cretaceous *Yaadia* spp. are more like those of epifaunal mytilids (Stanley, 1970, p. 27, text fig. 8), but those of Late Cretaceous *Yaadia* spp. (text fig. 9) are closer to those of infaunal and semi-infaunal mytilids. Stanley (1968, p. 220, and text fig. 2) classes all Trigoniacea as infaunal, but in addition to the blunt anterior profile *Yaadia* has large teeth with interlocking arcuate striae that prevent rocking of the valves. The looser fit of most heterodont hinge teeth permits such movements which aid burrowing. If an epifaunal *Yaadia* were disrupted by currents, its dentition would hold the valves closed with little energy expenditure until the clam was ready to put its foot out and reassume living position. The retentive clasp of these teeth held valves together in deposits in which nearly all of the other bivalves are found as single valves. Apparently the ligament was not strong enough to open the thick valves without the aid of the foot and thus would not have been an efficient aid to burrowing.

Newell and Boyd (1975) infer that Late Paleozoic trigonaceans were burrowers. These are predominantly unornamented, usually nearly equant bivalves having schizodian to myophorian grade dentition. Radiate and annulate sculpture first appear in the Permian (Newell and Boyd, 1975, p. 87). The acquisition of strong sculpture is roughly contemporaneous with the evolution of the trigonian grade dentition. Radi-

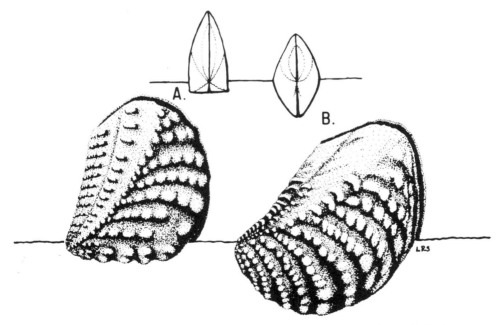

Fig. 8. Proposed living positions of a) *Yaadia lewisagassizi* and b) *Y. hemphilli.* Optimum effect of the sculpture in modifying water flow is obtained when the current moves from right to left.

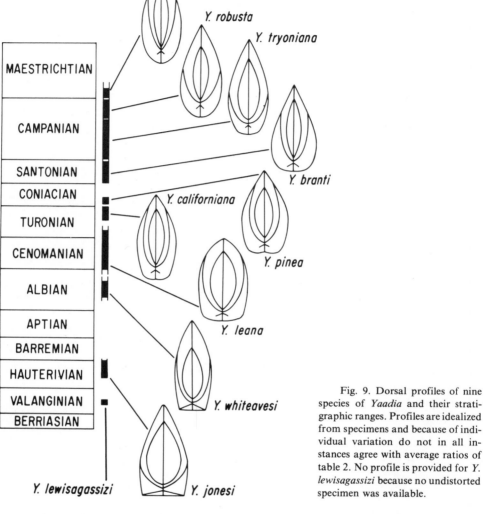

Fig. 9. Dorsal profiles of nine species of *Yaadia* and their stratigraphic ranges. Profiles are idealized from specimens and because of individual variation do not in all instances agree with average ratios of table 2. No profile is provided for *Y. lewisagassizi* because no undistorted specimen was available.

ate sculpture is eclipsed during the trigoniid heyday (Jurassic and Cretaceous) and reappears in the Late Tertiary. Radiately sculptured *Neotrigonia* is a shallow burrower, externally more similar to plump, rounded, radially sculptured cardiids which are also shallow burrowers. But this is not the shape and sculpture that Lycett (1872, p. 1) indicated as being so distinctive and well known that a description would be superfluous. The sculpture referred to by Lycett includes the oblique noded ribs of the myophorelline trigoniids. Such non-growth conforming sculpture has been considered to have strong functional significance (Seilacher, 1972, p. 325). If a trigoniid of myophorelline aspect, such as post-Turonian *Yaadia* spp. is placed foot down and inhalant-exhalant region up, the oblique flank ribbing is effectively radial along the anterior margin and anteroventral corner where it would most likely be useful in sawing into the substrate (text fig. 8). The ribs then arch upward toward the marginal angulation and at their dorsal ends are nearly parallel to the outer area valve margin. This parallel relationship between the dorsal portion of the flank ribs and the inhalant region seems—especially when present in Early Cretaceous *Yaadia* having a blunt anterior—more readily related to current-controlling than to sediment-burrowing functions.

Specimens of *Yaadia* spp. were placed in an aquarium, and water flow was generated by a pump. The specimens were stable when placed in sand that just covered the beaks with the shell margin of the outer area parallel to the sand surface. When the clam's sagittal plane is parallel to the current flow and the ventral margin directed upstream, the noded ribs act as riffles, disturbing the lamellar current flow next to the shell. Particles dance up the rib interspaces in the turbulence thus created. Flow across the inner area is turbulent but with a strong downward trend. Particles near the shell margin of the inner area are often whisked abruptly away downstream. Ribs and nodes of *Quadratotrigonia deslongchampsii* (Munier-Chalmas) and *Steinmanella transitoria quintocoensis* (Weaver) create similar small turbulent effects. The smooth surface of clams such as *Tivela stultorum* (Mawe) and *Callista (Megapitaria) aurantica* (Sowerby) do not cause such upward movement of particles along the flank, nor does the concentric ribbing of *Callista (Costacallista) erycina* (Linnaeus). Blue dye injected just upstream to and below the mid-point of the ventral margin of *Yaadia* positioned as in text figure 8 moved turbulently mainly up and toward the margins of the outer area and then down along the inner area and escutcheon. Sediment settled on the corcelet near the beaks of *Y. hemphilli* positioned as in text figure 8.

The hydrodynamic effect of sculpture is not yet well explained, but recent studies of ammonite sculpture (Chamberlain and Westermann, 1976; Chamberlain, 1976) have shown that like the indentations on a golf ball, sculpture does under certain conditions reduce drag. Rough concurrence in the development of the tight-fitting precise trigonian grade dentition and increase in shell ornamentation and size point toward a hydrodynamic function for the sculpture, and the nodded ribbs of the Myophorellinae may have evolved primarily to reduce drag on an uncemented, non-byssiferous epifaunal clam. Upward direction of movement of particles along the inter-ribs, however, may indicate some connection between this flow and the inhalant current. Even minor currents that would move food to the vicinity of the inhalant region and wastes away from the exhalant region could have been advantageous to these animals.

Text figure 8 shows *Y. lewisagassizi* with its beaks covered to the same extent as those of *Y. hemphilli* and positioned with the outer area shell margin parallel to the substrate. This results in a more exposed position for *Y. lewisagassizi*. Its blunt anterior, broadened by a ruff of large nodes, forms a base and the sediment-water interface roughly coincides with the unribbed zone anterior to the oblique flank ribs. Deecke (1926, p. 71) suggested that trigoniids in similar position used the friction of the knobby ribs to counteract the effect of gravity so that the inhalant-exhalant region could remain above the sediment surface. But an epi- to semi-infaunal trigoniid would need sufficient whole-animal density to maintain such a position. The considerable shell thickness of *Yaadia* (commonly 8 mm. in *Y. whiteavesi*) served to increase rather than counteract the effect of gravity. The thick shell and rugose sculpture, which would have made burrowing more difficult, would have stabilized the position of a recliner (Stanley, 1970, p. 35).

The posterior margin of the corcelet of *Neotrigonia* is nearly straight, but in *Yaadia* the margin of the outer area is at an angle of less than 180° to that of the inner area. The angle varies, but in general Early Cretaceous *Yaadia* have a more bent corcelet margin (angle less than 150°) than do Late Cretaceous *Yaadia* (angle greater than 150° but less than 170°). If, emulating *N. gemma,* a *Yaadia* is buried so that the outer area margin is at the sediment-water interface, the margin of the inner area is below the surface, requiring more effort to exhale without contaminating the inhalant current. With the corcelet margins above the substrate and the outer margin parallel to it, the inhalant area is uncontaminated by the exhalant. Fossil trigoniids with angled posterior corcelet margins as in *Yaadia* probably did not live buried to the valve margins but were instead epi- to semi-infaunal.

In their youth *Yaadia* spp. are square of outline, but as they mature the ventral and posterior margins grow fastest and the shells become elongate posteriorly (text figs. 11, 13a, 16, 19; tables 3-12; pl. 10, fig. 2-4, 7; pl. 11, fig. 3). Also in maturity the corcelet covers proportionately less of the shell surface. This ontogenetic growth pattern is repeated phylogenetically; in geologically older species the corcelet occupies more of the shell surface and in geologically younger species the marginal angulation migrates toward the dorsal margin. The greatest elongation of the shell which is dorsal to the marginal angulation is thus placed closer to the dorsal margin and the ventral shell margin is increased in length. Possibly this shortening of the marginal length of the corcelet (relative to the length of the ventral margin) (tables 3-12 C/H) is an indication that as the *Yaadia* lineage evolved there was an increase in the ability of the clams to pump and filter water. Numerous studies (See Wilbur and Owen, 1964, p. 231-232) have shown that the rate of filtration and the efficiency of feeding of bivalves decreases as the individual becomes larger. Hall (1975, p. 169) found that older individuals of *Tivela* and *Callista* do not deposit as many daily shell growth increments and the increments are not as thick. Studies (Wilbur and Owen, 1964, p. 229) show larger individuals to have grown more slowly and indicate that rate of growth may depend on size as well as age of individual. By rapid early growth individuals can soon be within the optimum size range for their habits and habitat, but sustaining such a growth rate would result in unfit size. Possibly those functions which would sustain the growth have been subject to fitting selection. The changes in the area-flank proportions of *Yaadia* spp. then could be related to increased ability to filter and feed through time and at any given time the necessary adjusting of these activities to produce an optimum-sized individual.

Increased ability to pump and filter water would be advantageous in an infaunal habit, but the cumulative suggestions of shape, sculpture, shell thickness, etc., are that adult *Yaadia* spp. were not buried to their posterior valve margins. The progressive changes in anterior nodes and shape, position of maximum width of shell, and proportion of corcelet to flank suggest that *Yaadia* slowly descended from epifaunal to semi-infaunal during the Cretaceous.

Yaadia spp. were conservative. They inhabited the same biotope and were generally accompanied by the same genera throughout the Cretaceous. They show limited evolution, changing only enough to adopt a more infaunal habit. They may have used the K-strategies of long life and limited reproduction. Growth checks are seldom apparent on *Yaadia,* and the shells are usually recrystallized; determining the chrono-logical age of any individual is thus difficult. Kauffman and Sohl (1974, p. 404) suggest that the open cellular wall structure of some large rudists was developed for rapid growth, but the shell structure of *Yaadia* is dense and provides no evidence of rapid growth. If *Yaadia* spp. had growth rates similar to *Litschkovitrigonia*? *fitchi,* the number of probable growth cehcks on *L.*? *fitchi* suggest an age in excess of 10 years for a large *Yaadia.* On the basis of egg size, Kauffman (1975, p. 167) has assigned bivalves to three groups: those producing very many small eggs; those producing a moderate number of large eggs; and those producing few very large eggs. If *Yaadia,* like *Neotrigonia,* produced large-sized eggs, that and the northern locus of its distri-bution would place it in the category of bivalves whose survival depends upon a moderate number of briefly planktonic larvae. The geologic record shows *Yaadia* to have been durably successful on or in the sandy bottoms near the North Pacific shores where *Yaadia* were probably equilibrium species. There is no apparent physical cause for the adoption by the genus of a more infaunal habit.

Biotic rather than physical factors seem more likely to have encouraged *Yaadia* (and other trigoniids) to seek the shelter of the substrate. Gastropods were becoming more capable carnivores and the clam-slugging stomatopods were already on the prowl. Whatever may have been the pressures encouraging burrowing, the lineage dissappeared as it developed a more infaunal habit. The more infaunal habit may have placed slowly changing, filibranch, asiphonate *Yaadia* into too direct competition with the more efficient venerids and mactrids which were evolving rapidly during the Cretaceous (Casey, 1952; Saul, 1973).

To the parallel evolutionary trends discussed by Newell and Boyd (1975) can be added: acquisition of drag reducing sculpture through the Triassic as the trigoniids became epifaunal; sculptural modifications such as the posteriorward progression of the V in forms assigned to *Vaugonia* (Jurassic) through *Apiotrigonia* (Cretaceous) into *Rutitrigonia* reflecting a return to more infaunal habit by several lineages; narrowing of the corcelet in most lineages from Jurassic through Cretaceous (e.g., *Scaphotrigonia* to *Pterotrigonia*). The change from oblique flank sculpture of *Eotri-gonia* to radial sculpture in *Neotrigonia* (Fleming, 1964, p. 196; but not Saveliev, 1958, p. 92, or Newell and Boyd, 1975, p. 67) may record another parallel movement to infaunal living. The success of *Neotrigonia* may result from the development of pallial ridges which would provide more flexible current control than the internal median ridges so prevalent among Mesozoic trigoniids, but the large and active foot that makes *Neotrigonia* so capable a burrower probably evolved in the early Mesozoic as a means of maintaining an upright position. The need of epifaunal trigoniids to evade predators doubtless encouraged development of the leaping foot.

Systematic Paleontology

Yaadia Crickmay, 1930
Type-Species by original designation:
Yaadia lewisagassizi Crickmay

Yaadia has generally been ignored or marginally commented upon by those study-ing the Trigoniidae, a neglect caused in large part by the preservation of the holotype of *Y. lewisagassizi*. It has been difficult to envisage from the published figures and Crickmay's brief description what *Y. lewisagassizi* looks like, whereas *Steinmanella* Crickmay, 1930b, and *Quadratotrigonia,* Dietrich, 1933, were based upon species represented by better preserved specimens from more accessible localities. Although tectonism has rendered the holotype of *Y. lewisagassizi* unphotogenic, those workers who have seen the holotype have recognized the distinctness of the genus, and Canadian paleontologists (McLearn *in* Sutherland Brown, 1968; Jeletsky *in* Coates, 1974) include *Yaadia* spp. on faunal lists. Neither Cox's (1952) use of *Yaadia* with *Steinmanella, Transitrigonia* Dietrich, 1933, and *Quadratotrigonia* as synonyms nor his later revision (*in* Moore, 1969), in which he recognized the genera *Steinmanella* and *Yaadia,* were popular with other workers. Under *Steinmanella* he grouped the subgenera *Litschkovitrigonia* Saveliev, 1958, *Setotrigonia* Kobayashi and Amano, 1955, and *Yeharella* Kobayashi and Amano, 1955; and under *Yaadia* he placed the subgenera *Leptotrigonia* Saveliev, 1958, and *Quadratotrigonia.* Some of the West Coast species herein assigned to *Yaadia* were so assigned by Cox. Packard's (1921) grounds for placing *Trigonia california* Packard, *T. tryoniana* Gabb, and *T. jack-sonensis* in the Glabrae group and *T. leana, T. l. whiteavesi* Packard, and *T. fitchi* Packard in the Quadratae group are unclear to me. Japanese authors have referred the above, excepting *T. jacksonensis* which resembles *Rutitrigonia,* to *Steinmanella (Yeharella),* but *T. fitchi* is morphologically closer to *Litschkovitrigonia.* Saveliev (1958, p. 105) has also used the subgenus *Yeharella* for these species, but he has placed them in the genus *Quadratotrigonia.*

The tectonically distorted holotype of *Y. lewisagassizi* is a left valve from Harrison Lake, British Columbia. The most distinctive feature recognized by Crickmay was the "discrepant ornament on the anterior end of the disk"; and because of this remarkable ornament he separated *Yaadia* from similarly quadrate, nodose *Steinmanella.* Addi-tional scrappy topotype impressions include the beak portion of a right valve, but they add little to the definition of the genus because of the distortion and type of preser-vation—casts in a coarse-grained sandstone. Less distorted material ranging in age from Hauterivian to Maestrichtian indicates that the genus *Yaadia* is character-istically a large, thick-shelled trigoniid of quadrate outline (table 2). The beak is near the anterior end of the shell and the posterior is elongate comprising five-sixths or more of the shell length. In dorsal profile *Yaadia* is strongly inflated anteriorly, tapering posteriorly (text fig. 9). The zone of maximum inflation of the valve at any growth stage is along or near the marginal angulation. Juvenile ribs are subconcentric and continuous from anterior to posterior margin but usually with a jog at the

TABLE 2

Averages of ratios of measurements for 9 species of *Yaadia* (based on tables 3-11)

	H/L	T/H	C/H
jonesi	.74 (.71)	.26 (.24)	.63 (.62)
whiteavesi	.71 (.72)	.25 (.26)	.6 (.6)
leana	.81 (.8)	.32 (.33)	.68 (.67)
californiana	.76 (.72)	.27 (.25)	.58 (.57)
pinea	.7 (.71)	.32 (.33)	.57 (.55)
branti	.75 (.7)	.35 (.34)	.56 (.55)
tryoniana	.75 (.68)	.31 (.32)	.56 (.53)
robusta	(.75)	(.33)	(.53)
hemphilli	.75 (.73)	.32 (.32)	.52 (.52)

H = height in sagittal plane perpendicular to ligament; L = length parallel to ligament; T = thickness of one valve; C = length of posterior margin of corcelet. Numbers in parentheses are averages of ratios of specimens with H 50 mm. or greater. *Y. leana* is mainly represented by medium sized specimens; if the ratios of those with H greater than 60 mm. are averaged, the ratios are H/L = .79, T/H = .34, C/H = .66. The least difference between species is in the H/L ratio. All species are more equant in their youth and become more elongate in maturity. According to the T/H, *Yaadia* becomes plumper through time. C/H is not a measurement of the area of the corcelet relative to the area of the flank, but it gives an indication of proportionate increase or decrease of these areas. Smaller shells usually have a broader corcelet area and so do geologically older species. More mature shells generally have a narrower corcelet area and so do geologically younger species.

marginal angulation and an emargination at the median groove. Adult ornament is separated into discrete units. The flank is strongly ornamented with broad oblique ribs embossed with nodes. The ornament of the anterior slope is a specific characteristic. The marginal angulation is marked by a row of nodes generally until the shell attains 50 mm. in length. On the outer area, bordering the median groove is a row of nodes which evanesces when the shell is between 40 and 50 mm. long. The inner area also has a row of nodes, but its placement varies as to species being either along the median groove or bordering the escutcheon. These nodes are often elongate, and almost rib-like. The remarkable discrepant sculpture noted by Crickmay is found only on Early Turonian and older species. It consists of a row of very large nodes along the anterior angulation. In the oldest species the flank ribs are separated from this anterior set of nodes by a smooth area. In geologically younger species the nodes become more similar in size to the nodes on the flank ribs and they and the flank ribs are less and less set apart. In specimens of Early Turonian age the anterior set of nodes may best be recognized by their ill-alignment with the flank ribs. Species of Late Turonian and younger age do not have a separate row of nodes. Along the ventral margins (and anterior margins in those species having ribs to the anterior margin edge) the ribs of the right valve alternate with those of the left valve. As a result the sculpture on the two valves of one individual differs as to number of flank ribs, placement of flank ribs, and nodes per rib. A rough estimate of node size can be made by counting the nodes on the rib which runs from the marginal angulation to the ventral end of the anterior margin (see text fig. 1). This is the "corner rib." In those species which have a row of distinctive nodes along the anterior angulation, the corner rib is the first flank rib to extend to the valve margin.

The sculpture of *Yaadia* is typically untidy. The nodes are roughly graded as to size, with those nearest the beak and the marginal angulation being smallest, but adjacent nodes are often irregularly larger or smaller. In some species the nodes tend to coalesce along the rib; in others the nodes tend to coalesce across the interspace following the growth line and form concentric cords usually near the ventral margins (pl. 6, fig. 6; pl. 7, fig. 9; pl. 11, fig. 3). A similar sculpture development takes place in species from Japan assigned to *Steinmanella (Yeharella)* by Kobayashi and Amano (1955, p. 197): *S. (Y.) japonica* (Yehara) and *S. (Y.) kimurai* (Tokunaga and Shimizu). Well-preserved specimens of *Yaadia* show fine wrinkling of the surface parallel to the growth lines. The wrinkles are interrupted or deflected by the nodes which they usually do not override. The shell layer making up the wrinkles is very thin and readily lost.

Inside, the right valve has two long, ridged cardinal teeth and the left valve has three cardinals. The large middle tooth of the left valve is clearly but not deeply bifid. This was obviously a precisely fitting hinge; the ridges on the teeth are arcuate, reflecting the arc of the valve opening. Just posterior to the posterior adductor muscle scar on the broad shell margin are low, short lateral teeth and their sockets set at an angle near 40° to the posterior valve margin; 3 teeth in the right valve and two in the left valve. The muscle scars are clearly marked. The adductors are close to the hinge, the posterior a little larger than the anterior. The anterior adductor and pedal retractor muscle scars are on a heavy myophoric buttress immediately adjacent to the cardinal teeth. The elevator pedal muscle scar is in a deep pit under the large middle tooth of the left valve and between the two cardinals of the right valve. The pallial line is entire but distant from the valve margins. The interior of the corcelet is marked by two ridges, a low ridge along the marginal angulation and a stronger ridge along the median groove.

Yaadia differs from *Steinmanella* in having smoother inner and outer areas. Both areas may be and usually are concentrically ribbed very near the beak, but they are relatively smooth on adult shells and lack the growth-line oriented rugosities which develop on adult *Steinmanella*. Early Cretaceous *Steinmanella* lacks the distinctive row of large nodes on the anterior angulation so prominent on *Yaadia* of similar age and the three rows of prominent nodes marking the marginal angulation, the median groove, and the escutcheon. Two specimens of *Steinmanella transitoria quintucoensis* (Weaver) from the Weaver collection at UCR have a sulcus in the growth lines just anterior to the marginal angulation (pl. 11, fig. 7). No specimen of *Yaadia* has such a sulcus. The rounder anterior profile of Late Cretaceous *Yaadia* is similar to *Steinmanella* but by that time the marginal angulation and median groove of *Yaadia* are gently concavely curved to the posterior valve margin rather than straight. This curvature is similar to that of *Quadratotrigonia* which Cox (*in* Moore, 1969, p. N488) included as a subgenus of *Yaadia*. *Yaadia,* however, does not have the distinctive pustular ornament of the inner and outer area of typical *Quadratotrigonia*. *Quadratotrigonia* is more equant; *Yaadia* is elongate posteriorly with more anteriorly placed beaks and is more strongly inflated anteriorly. As in the case of *Steinmanella,* Early Cretaceous *Quadratotrigonia* lacks the distinctive row of clearly set apart large nodes on the anterior angulation characteristic of Early Cretaceous *Yaadia*. *Leptotrigonia* Saveliev, 1958, also included as a subgenus of *Yaadia* by Cox (*in* Moore, 1969) lacks the anterior row of nodes and has more irregular sculpture on the area. The adult elongate rugosities of the area are more suggestive of *Steinmanella* than *Yaadia*. *Leptotrigonia* is considered a synonym of *Quadratotrigonia* by Nakano (1968, p. 35). Conversely, species of *Yeharella* Kobayashi and Amano (1955, p. 200), described as a subgenus of *Steinmanella,* are very similar to *Yaadia* spp. of equivalent age. *Setotrigonia* Kobayashi and Amano (1955, p. 206), also described as a subgenus of *Steinmanella* is more like a *Quadratotrigonia* in shape as is *Litschkovitrigonia* Saveliev (1958, p. 97).

In addition to *Yeharella* (type-species *Trigonia japonica* Yehara), Kobayashi and Amano (1955, p. 197) introduced the name *Packardella* for the North American group of species which they assigned to *Steinmanella: Trigonia californiana* Packard, *T. tryoniana* Gabb, *T. fitchi* Packard, *T. leana* Gabb, and *T. whiteavesi* Packard. *T. fitchi* is unrelated to the other species of this group. It is not a *Yaadia,* and its resemblances are to *Litschkovitrigonia. T. whiteavesi* has well-developed discrepant anterior sculpture similar to that of *Yaadia lewisagassizi,* and this sculpture is still apparent on *T. leana. T. cf. T. californiana* and *T. tryoniana* are without such sculpture and either could typify a group resembling *Yeharella* and distinguishable from *Yaadia,* but both species are based on poor material. Applying a subgeneric name to post-Mid-Turonian *Yaadia* spp. also tends to obscure their direct relationship to *Yaadia.* Kobayashi and Amano did not name a type-species for *Packardella;* the name is thus unavailable and is considered best left so.

Yaadia resembles *Myophorella* in having well-developed subumbonal elevator pedal muscle scars, especially in the left valve (Fleming, 1964, p. 198) (pl. 1, fig. 6; pl. 3, fig. 6) and in the oblique noded ribs of the flank. It differs from typical *Myophorella* in having a noded marginal angulation, nodes along the median groove and nodes along the escutcheonal angulation. The escutcheonal plane of *Yaadia* is nearly normal to the sagittal plane and the escutcheon is usually ornamented by strings of nodes or slightly nodular ribs.

Crickmay (1930b, p. 50) stated that the similar anterior ornamentation of *Scaphogonia, (Scaphitrigon* is a *lapsus calami, fide* Crickmay, 1962, p. 11), was not necessarily an indication of relationship. *Scaphogonia argo* Crickmay differs from *Yaadia* spp. in the concave rather than convex inflation of the area, in the more angulate marginal angulation, weaker nodes along the median groove and escutcheonal angulation, and having an escutcheon more like that of *Myophorella.* If *Scaphogonia* is ancestral to *Yaadia,* these differences should become less noticeable in Late Jurassic *Scaphogonia* spp.

Yaadia lewisagassizi Crickmay

(Text figs. 8, 10)

Yaadia lewisagassizi Crickmay, 1930b, p. 50, pl. 13, fig. 1-2; Saveliev, 1958, p. 106, pl. 54, fig. 2a-b; Bolton, 1965, p. 183; Cox *in* Moore, 1969, p. N488, fig. D74, 4.

Description.—Owing to tectonic distortion, shape cannot be described accurately, but the species appears to have had an abruptly truncated anterior end with a strong anterior angulation from beak to ventral margin. The beak probably very near the anterior end of the shell; zone of maximum inflation along the marginal angulation. Anterior angulation made more prominent by a double row of very large nodes, from each node a string of smaller diminishing nodes descends the steep slope to the anterior margin. Nodes of the anterior angulation separated from the oblique flank ribs by an unornamented space wider than any flank rib interspace. Nodes of the flank ribs large, corner rib bearing about 7. Oblique ribs covering the flank even to the posterior end. Marginal angulation bearing a row of large nodes, also to the posterior end. Row of nodes on the outer area bordering the median groove unusually large and persisting to the posterior margin. Median groove very well marked. Nodes of inner area elongate, bordering the escutcheon. Escutcheon intricately ornamented with rows of nodes corresponding to and extending from the nodes of the inner area to the dorsal margin.

Holotype.—CGS cat. no. 9702.

Topotypes.—UCLA cat. no. 49637.

Type locality.—"South shore of a little point on the west side of Cascade Bay, 1,090 yards south of the north end of that bay, the Peninsula, east side Harrison Lake, British Columbia" (loc. 47 of Crickmay 1930b and 1962 = UCLA loc. 6177).

Age.—Mid Valanginian ?

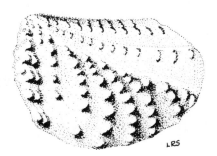

Fig. 10. Reconstruction of *Yaadia lewisagas-sizi* Crickmay using average height/length ratio of *Y. jonesi* as basis for shape.

Remarks.—Dr. T. E. Bolton of the Geological Survey of Canada has very kindly provided me with a rubber pull (UCLA cat. no. 49639) of the holotype of *Y. lewisagassizi*. My measurements of height, 56 mm.; length, 90 mm.; and thickness, 18 mm. made on this pull do not agree with Crickmay's of height, 63 mm.; length, 77 mm.; thickness, 22 mm. It seems likely that the measurements were taken in different positions. The height/length ratio from Crickmay's measurements is .81 and from mine it is .62. Measurements of 86 specimens of *Yaadia* (tables 2-11) give an overall H/L average of .74, and the average H/L for *Y. jonesi*, the species most similar to *Y. lewisagassizi*, is .73. The reconstruction of *Y. lewisagassizi* (text fig. 10) is based on this ratio. The ornament is derived from rubber pulls of the holotype and two topotype scraps (UCLA cat. no. 49637). This approximation of *Y. lewisagassizi* before stretching should be used with discretion; the figure provides a visual aid for recognizing the genus *Yaadia*, but the only feature that can be indicated as charac- terizing the species is the strong double row of nodes along the anterior angulation. *Yaadia* of younger age have a single prominent row of nodes along the angulation. More specimens of *Y. lewisagassizi* are needed to determine its range of variation and other differences between it and *Y. jonesi*.

Crickmay (1930b, p. 42, 49, 51) listed *Aucella teutoburgensis* Weerth from the type locality of *Y. lewisagassizi*. He later (1962, tabular summary of sections) clearly placed this locality at the top of the Brokenback Hill Fm. Jeletzky (1965, p. 49) implied some doubt as to the Hauterivian age of the *Yaadia* beds based on the occurrence of *Buchia teutoburgensis* indicating that these were the only Hauterivian *Buchia* from the "whole of the western Cordilleran belt of North America." The holotype of *Y. lewisagassizi* still retains Crickmay's original field no. 269 (Bolton, 1974, letter); the same number is on the UCLA *Yaadia* scraps and on two specimens of *Buchia* sp.

(UCLA cat. no. 49636). These are evidently the *Aucella teutoburgensis* of Crickmay. Both D. L. Jones and J. A. Jeletsky have looked at these specimens and concur that they are of mid Valanginian age. Jeletsky mentioned the possibility of careless collecting or subsequent mixing of fossils. Thin sections of (1) rock labeled 269, bearing *Yaadia* fragments; (2) a scrap extracted from the larger *Buchia* specimen; and (3) matrix attached to the smaller *Buchia* were shown to Dr. A. G. Barrows, California Division of Mines and Geology. He found that the *Yaadia* matrix is not a "bright green tuff" (Crickmay, 1962, p. 9) but a greenschist that was originally a basalt-derived, volcano-clastic graywacke. The scrap of matrix from the larger *Buchia* is a similar rock, but the matrix of the smaller *Buchia* is quartz rich and much better sorted. The age, then, of *Yaadia lewisagassizi* is still in doubt, but what evidence there is indicates a mid Valanginian age.

Jeletsky (*in* Coates, 1974, p. 11, 29, 86, 96, 129, 131, 151, and table 8) identified *Yaadia lewisagassizi* in degrees of certainty and *Yaadia* sp. from Barremian age beds in the Jackass Mountain Group (Map Unit 8) of the Manning Park area, British Columbia. I have not seen these specimens and do not know how they compare with *Y. lewisagassizi* from Harrison Lake area or *Y. jonesi* of Hauterivian age from the Days Creek Formation, Southern Oregon.

Yaadia jonesi n. sp.
(Pl. 1, figs. 1-8; pl. 2, figs. 1-2; pl. 11, fig. 1; text fig. 11; table 3)

Description.—Dorsal margin nearly straight. Anterior end abruptly truncated with a strong anterior angulation from beak to ventral margin. Beak very near the anterior end, nearly terminal. Zone of inflation broad, from marginal angulation to second n de of flank rib. Anterior angulation accentuated by row of very large nodes from each of which a riblet decends, dying out toward the anterior margin. Unornamented zone flankward of the anterior nodes less wide than flank rib interspaces. Oblique flank ribs narrower than interspaces, covering the flank to the posterior end. Corner rib bearing about 8 nodes. Marginal angulation nodes nearly as large as flank rib nodes, continuing to or nearly to posterior end. Nodes of outer area bordering the median groove, smaller than nodes on angulation, dying out near posterior end. Median groove very well marked. Nodes of inner area elongate, bordering the escutcheon, strongly accenting an escutcheonal angulation. Escutcheon broad, intricately ornamented with rows of small nodes corresponding to and extending from the nodes of the inner area to the dorsal margin. Escutcheonal nodes sporadically coalesced along growth lines.

Syntypes.—UCLA cat. nos. 38567 and 38568.

Paratypes.—UCLA cat. no. 38566 from UCLA 6275; USNM cat. no. 241670 from USGS loc. 22498, USNM cat. nos. 241671 and 241672 from USGS loc. 1245; plaster cast of uncollected mold CAS cat. no. 57982 from CAS loc. 33767.

Hypotype.—USNM cat. no. 241669 from USGS loc. 3339.

Dimensions.—See table 3.

Type locality.—UCLA loc. 6276, South Umpqua River just north of Days Creek, Douglas Co., Oregon.

Age.—Mid to late Hauterivian, ?Early Barremian, *Hollisites dichotomus* and *Hertleinites aguila* zones.

Remarks.—This species is based upon six specimens from near the mouth of Days Creek on the South Umpqua River, four from near Riddle, Douglas Co., and one plus fragments from the Logan Cut approximately 4 miles south of Cave Junction, Josephine Co., Oregon. Imlay (1960, p. 177) has referred the Days Creek and Riddle localities to the *Hollisites dichotomus* zone of mid Hauterivian age and (1960, p. 180) the Logan Cut localities to the *Hertleinites aguila* zone of late Hauterivian age. The distortion of specimens of *Y. lewisagassizi* makes uninterpreted comparison with any other species impossible. An undistorted *Y. lewisagassizi* probably looked like text

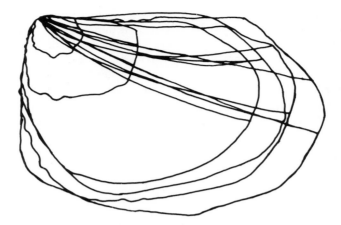

Fig. 11. Diagrammatic outlines of *Yaadia jonesi* Saul
superposed with beak and anterior margins matching as much
as possible. CAS cat. no. 57982, UCLA cat. nos. 38566,
38567+38568 (compound reconstruction), USNM cat. nos.
241669-241671. All drawn as left valves, margins restored, and
large specimens reduced about one third.

TABLE 3

Measurements (in mm.) of specimens of *Yaadia jonesi*

	loc.	H	L	T	C	H/L	T/H	C/H	Remarks
38566	6275	12	14	4	9	.86	.33	.75	rubber pull
241669	3339	24	30	7	13	.8	.29	.54	rubber pull
57982	33767	61	88	16	39	.69	.21	.64	plaster cast
241671	1245	70	97	20	40	.72	.28	.57	broken
241675	1245	71?	105?	20	41?	.68	.28	.58	broken
38567 38568	6276	81	106	20	50?	.77	.24	.62	composite rubber pulls
241670	22498	81	120?	25	57	.67	.21	.7	broken
averages of ratios						.74	.26	.63	
averages of ratios H = 50 mm. +						.71	.24	.62	

H = height sagittal plane perpendicular to ligament. L = length in sagittal plane parallel to ligament.
T = thickness of one valve. C = length posterior margin of corcelet.

fig. 10, and the species is very similar to *Y. jonesi*. *Y. jonesi* has a single rather than
double row of strong nodes along the anterior angulation and the ribblets on the
anterior face are only slightly nodular. Judged from the latex pulls, on *Y. lewisagassizi*
these riblets are more nodose, more prominent and probably extend to the anterior
margin. The nodes on the flank ribs appear larger in *Y. lewisagassizi* than in *Y. jonesi;*
an appearance that may be due to the types of preservation, but it is supported by *Y.
lewisagassizi* having 7 rather than 8 nodes on the corner rib. Given the small number
of specimens of each of these species—one *Y. lewisagassizi,* three *Y. jonesi*—upon

which the corner rib nodes can be counted and the variability in *Yaadia* ornamentation, this difference must be considered to be in need of verification.

The species is named for D. L. Jones who generously contributed many specimens to this study.

Yaadia whiteavesi (Packard)

(Pl. 2, figs. 3-5; pl. 3, figs. 1-6; text fig. 12, table 4)

Trigonia, sp. indt., Whiteaves, 1876, p. 70, pl. 10, fig. 2, 2a.
Trigonia leana var. *whiteavesi* Packard, 1921, p. 21, pl. 6, fig. 2. (not pl. 5, fig. 4 = *Y. hemphilli* (Anderson));
    Stewart, 1930, p. 93
Trigonia perrinsmithi Anderson, 1958, p. 110, pl. 2, fig. 7.
Trigonia whiteavesi (Packard) Anderson, 1958, p. 111.

Description.—Dorsal margin nearly straight. Anterior end abruptly truncated with a strong anterior angulation from beak to ventral margin. Beak very near anterior end, nearly terminal. Zone of inflation along first node of flank ribs. Anterior angulation accentuated by row of very large nodes. Median nodes of row with ventrally deflected slightly nodular riblets evanescing about halfway to anterior border. Nodes of more ventral riblets, larger, coalescing along growth lines. Unornamented interspace flankward of the anterior nodes about one-half as wide as flank rib interspaces. Oblique flank ribs and interspaces about equal in width, covering the flank to the posterior end in medium-sized specimens, fading near the ventral margin-marginal angulation of large specimens. Corner rib bearing about 9 nodes. Nodes roundish with crests slightly elongated in a growth line direction. Marginal angulation nodes half as large as flank nodes, dying out 60-70 mm. from beak. Nodes of outer area smaller than marginal angulation nodes, elongate, bordering the well-marked median groove, dying out 50-60 mm. from beak. Inner area with fine riblets near beak becoming shorter and lying along escutcheonal angulation, larger and more irregular near posterodorsal margin. Escutcheon nearly smooth with scattered small nodes.

Holotype.—CGS cat. no. 4997 (plaster cast UCLA cat. no. 32195).
Paratype.—CGS cat. no. 4997a (plaster cast UCLA cat. no. 32196).

TABLE 4

Measurements (in mm.) of specimens of *Yaadia whiteavesi* (Packard)

	loc.	H	L	T	C	H/L	T/H	C/H	Remarks
38575	4670	48	71	13	28	.68	.27	.58	rubber pull
38753	3909	50	67	12	30	.74	.24	.6	broken
38577	M253	54	72	15	37	.75	.27	.68	rubber pull of USNM 241674
38576	4670	56	79	16	30	.71	.28	.54	rubber pull
32195		71	88	16	42	.8	.22	.59	plaster cast of crushed? holotype posterior broken
38574	3915	82	120	20	?	.68	.24	?	no shell
48662		82	124	21	53	.66	.26	.64	plaster cast of holotype *T. perrinsmithi*
38570	3909	92	127	26	50	.72	.28	.54	broken
averages of ratios						.71	.25	.6	
averages of ratios H = 50 mm. +						.72	.26	.6	

H = height in sagittal plane perpendicular to ligament. L = length in sagittal plane parallel to ligament. T = thickness of one valve. C = length posterior margin of corcelet.

Hypotypes.—UCLA cat. nos. 38570, 38753 from UCLA loc. 3909; 38574 from UCLA loc. 3915; 38575-76, 38836 from UCLA loc. 4670; USNM cat. no. 241674 from USGS loc. M253 and 241673 from USGS loc. 1051; LSJU cat. no. 8697 (holotype of *T. perrinsmithi*).

Dimensions.—See table 4.

Type locality.—Queen Charlotte Islands.

Age.—Late Early Albian (*Brewericeras hulenense* zone) to middle Albian.

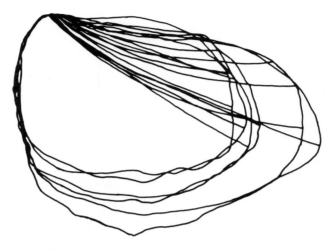

Fig. 12. Diagrammatic outlines of *Yaadia whiteavesi* (Packard) superposed with beak and anterior margins matching as much as possible. UCLA cat. nos. 32195, 38570, 38575-38577, 48662. All drawn as left valves, margins restored, and all reduced about one third.

Remarks.—The above description is based upon eight specimens from Texas Springs and one from Horsetown (the type of *T. perrinsmithi*), Shasta Co., California, several natural rock molds of specimens from Grave Creek, Jackson Co., Oregon, from five of which excellent and nearly complete latex pulls were made, and plaster casts of the holotype and paratype from Queen Charlotte Islands, British Columbia.

There is no possible basis for Anderson's (1958, p. 111) choice of CGS 4997a as holotype of this species. While it may, as Whiteaves said, "give the clearest idea of the normal shape of the shell," it is missing so much shell that it does not give any idea of the sculpture (pl. 2, fig. 5) and therefore provides much less data upon which to base the species than CGS 4997 (pl. 2, fig. 3). Whiteaves' figure of CGS 4997 is not a restoration or a composite as suggested by Anderson; the careful reading of Whiteaves urged by Anderson is beside the point as Packard is the author of the species. As noted by Stewart (1930, p. 93), Packard's species is based upon Whiteaves, 1876, pl. 10, fig, 2a. Packard refigured only CGS 4997, referred to the original figure (singular) and discussed the probable locality of the figured specimen (singular). CGS 4997a is here considered a paratype because Packard listed Whiteaves' figure of it in his synonymy and quoted Whiteaves' description which included reference to this specimen.

The type locality of *Y. whiteavesi* (Packard) is not specific. Specimens collected by James Richardson in 1872 (Whiteaves, 1876) were described as being from the Queen Charlotte Islands, British Columbia. Bolton (1965, p. 176), following Richardson's notations, lists the locality of the type specimens as west of Alliford Bay, Skidegate Inlet, Queen Charlotte Islands, in the Haida Formation. The Haida Formation, however, crops out only to the east of Alliford Bay, Moresby Island (Mackenzie, 1916, maps 176A and 177A; Sutherland Brown, 1968, fig. 5 [B]; McLearn, 1972, fig. 1), and Richardson's notation "west of Alliford Bay" means nothing more specific than Skidegate Inlet (McLearn, 1972, p. 44). Packard (1921, p. 22) favored the north side of Maude Island. F. H. McLearn and D. L. Jones identified fossils, including *Trigonia (Yaadia) leana* var. *whiteavesi* Packard from there, Bearskin Bay, and Lina Island as typical of the *Brewericeras hulenense* zone (Sutherland Brown, 1968, p. 92 and table 9), but I have no evidence suggesting which of the three probable is the actual type locality.

The largest specimens of this species, and of the genus, are from the Budden Canyon Formation, Texas Springs (UCLA loc. 3909; USGS loc. 1051) and Horsetown (the holotype of *T. perrinsmithi* Anderson), Shasta Co., California. A beak fragment from UCLA loc. 3909 has a posterior cardinal tooth 32 mm. long. Specimens from Grave Creek, Jackson Co., Oregon, are generally half as large as those from Shasta Co. Whiteaves' types are larger than most Grave Creek specimens but not more than two-thirds as large as some Shasta County specimens. The Grave Creek beds have been dated as late early to middle Albian (Jones, 1960a, p. 155) and these specimens may be younger than those from Skidegate Inlet and Texas Springs, but no stratigraphic significance is attached to these size differences nor does this necessarily indicate that the species grew larger in the southern part of its range.

Y. whiteavesi differs from *Y. jonesi* in having a narrower space between the anterior row of nodes and the flank ribs; flank ribs which are slightly wider relative to their interspaces, and more moderate sculpture on the corcelet.

Y. whiteavesi has a much more well-developed anterior row of nodes than does *Y. leana*.

Yaadia leana (Gabb)

(Pl. 4, figs. 1-8; pl. 5, figs. 1-4; pl. 11, figs. 2; text fig. 13a-b; table 5)

Trigonia Gibboniana Lea?, Gabb, 1864, p. 190, pl. 25, fig. 178; pl. 31, fig. 262.
Trigonia leana Gabb, 1877, p. 312 (new name for *T. gibboniana* of Gabb not Lea); Packard, 1921, p. 20, pl. 5, figs. 1-3, 5-6; pl. 6, fig. 1; pl. 7, fig. 1; Stewart, 1930, p. 92; Anderson, 1958, p. 113.
Trigonia californiana Packard, 1921, p. 18, "CAS specimen 12784" only.
Trigonia colusaensis Anderson, 1958, p. 110, pl. 1, fig. 6.
Trigonia wheelerensis Anderson, 1958, p. 116.

Description.—Dorsal margin nearly straight. Anterior end truncated but convex with high rounded anterior angulation. Beak very near anterior end. Zone of inflation along marginal angulation or first node of flank ribs. Anterior angulation accentuated by row of moderate nodes. Anterior slope roughened by growth lines, some specimens with a few nodes aligned with nodes of the anterior angulation. No well-marked interspace between anterior angulation nodes and oblique flank ribs. Flank ribs approaching anterior margin nearly equal in number to anterior angulation nodes. Oblique flank ribs and interspaces about equal in width, nearly straight, covering the flank to the posterior end. Corner rib with 6 or 7 nodes Nodes elongate along ribs with tendency to coalesce along ribs and along growth lines near antero-ventral margin. Marginal angulation nodes half as large as flank nodes, dying out 50-60 mm. from beak. Nodes of outer area smaller than marginal angulation nodes, bordering the median groove, dying out 40-50 mm. from beak. Outer area notably broad and smooth. Inner area narrower, crossed by narrow riblets near beak which continue to be about 5 mm. long and lie along and rougly normal to escutcheonal angulation. Escutcheon usually smooth, occassionally with scattering of small nodes near escutcheonal angulation.

  Lectotype.—UCBMP cat. no. 12171 (plaster cast UCLA cat. no. 28701).

  Paralectotypes.—UCBMP cat. nos. 14498-14499 from "Jacksonville."

  Hypotypes.—CAS cat. nos. 10632 (=CSMB 12784, holotype of *T. colusaensis*), 57980 from CAS loc. 29580; UO cat. nos. 26909 (holotype of *T. wheelerensis*) from UO loc. 92, 26897 from Jackson Co.; UCLA cat. nos. 38578-79 from UCLA loc. 3763, 38580-84 from UCLA loc. 6168, 38590-92 from UCLA loc. 6307, 38594 from UCLA 5425, 38596 from UCLA 5427, 38715-23 and 38754 from UCLA loc. 6306; USNM cat. nos. 241677-241678 from USGS loc. 28268, 241675-241676, 241819 from USGS loc. 10117, and 241679 from USGS loc. 2187.

  Dimensions.—See table 5.

  Type locality.—"Martinez," Contra Costa Co., California. (The matrix resembles that from Curry Canyon, south side Mt. Diablo.

  Age.—Cenomanian to early Turonian.

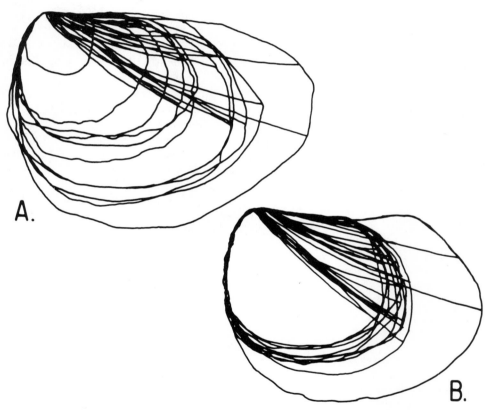

Fig. 13. Diagrammatic outlines of *Yaadia leana* (Gabb) superposed with beak and anterior margins matching as much as possible. All drawn as left valves, margins restored, and larger specimens reduced about one third.
  a. UCLA cat. nos. 28701, 38578-38580, 38582, 38596, 48655, 48660; USNM cat. nos. 241675, 241677, 241679.
  b. UCLA cat. nos. 38715-38723, 38754. All from UCLA loc. 6306, Bellinger Hill, Jackson Co., Oregon.

TABLE 5

Measurements (in mm.) of specimens of *Yaadia leana* (Gabb)

	loc.	H	L	T	C	H/L	T/H	C/H	Remarks
28701		18	20	5	11	.9	.28	.6	plaster cast of lectotype
241678	26268	34	42	15	25	.81	.44	.73	
241677	26268	38	44	10	24	.86	.26	.63	
38583	6168	41	53	14	29	.77	.34	.71	
38578	3763	42	49	12	29	.86	.28	.67	
38584	6168	42	50	12	27	.84	.28	.64	
38592	6307	42	52	12	31	.81	.29	.73	rubber pull
38596	5427	42	57	10	28	.74	.24	.67	
241675	10117	42	58	15	30	.72	.36	.71	
38723	6306	43	54	14	28	.8	.33	.65	
38717	6306	46	57	16	34	.81	.34	.74	
241819	10117	47	53	16	31	.88	.34	.66	broken
38716	6306	47	56	16	35	.84	.34	.74	
38782	6168	47	57	19	30	.82	.4	.64	
38590	6307	47	58	16	34	.81	.34	.72	broken
38594	5425	48	63	13	37	.76	.27	.77	broken
38722	6306	48	58	15	30	.83	.31	.62	
38721	6306	48	60	15	31	.8	.31	.65	
38720	6306	48	60	17	33	.8	.35	.68	
38581	6168	49	55	13	32	.89	.26	.65	
38591	6307	49	61	12	39	.8	.24	.79	broken
38719	6306	49	55	17	35	.89	.34	.71	
38715	6306	50	59	16	36	.85	.32	.72	
38718	6306	51	60	17	36	.85	.32	.71	
241679	2187	55	66	15	36	.83	.27	.65	
48655	92	57	79	20	38	.72	.35	.67	plaster cast of holotype *T. wheelerensis*
14498	A-6458	58	70	14	40	.83	.24	.68	lectoparatype eroded, T too low
38579	3763	59	76	19	?	.78	.32	?	no shell on corcelet and beak
38580	6168	60	67	22	37	.9	.37	.62	
48658		60	75	18	45	.8	.3	.75	plaster cast of hypotype UO 26867
38754	6306	62	84	22	48	.74	.35	.77	broken
48660		72	99	25	35?	.73	.35	.53?	plaster cast of holotype *T. colusaensis*, nearly shelless

averages of ratios						.81	.32	.68	
averages of ratios H = 50 mm. +						.8	.33	.67	
averages of ratios H = 60 mm. +						.79	.34	.66	

H = height in sagittal plane perpendicular to ligament. L = length in sagittal plane parallel to ligament. T = thickness of one valve. C = length posterior margin of corcelet.

Remarks.—The above description is based upon the lectotype and 4 scraps from UCB loc. 82 Curry Canyon; the paralectotypes from "Jacksonville," Oregon; 8 specimens from UCLA loc. 6186 Rock Creek, Dayville quad., Oregon; 7 specimens from USGS localities 6-7 miles south of Dayville, Dayville quad., Oregon; 20+ specimens from several localities near Jacksonville, Jackson Co., Oregon; 3 specimens from UCLA loc. 3763 Bald Hills Formation, Ono area (Murphy and Rodda, 1960, p. 856); and the holotypes of *T. colusaensis* and *T. wheelerensis.* Four scrappy specimens from CAS loc. 29580, 1 mile WNW of San Luis Ranch, Pacheco Pass quad., Merced Co., California are also probably *Y. leana,* as are two beak fragments and a medium-sized external mould of a *Yaadia* from USGS loc. M175 near Sites, Colusa Co., California. Abundant broken and peeled *Yaadia* which are probably *Y. leana* were collected from conglomerate beds on Elder Creek (UCB loc. 2604), Tehama Co., California. A poorly preserved beak fragment from the Ortigalita Peak quadrangle, about 1 mile southeast from Piedra Azul Spring (USGS loc. M6003), Merced Co., California is also probably *Y. leana.*

Specimens of *Yaadia* are common in the vicinity of Jacksonville, Oregon. Packard (1921) and Anderson (1958) considered these to be typical *T. leana* and of Turonian age. Some Jacksonville specimens can not be distinguished from Dayville specimens; others are less truncated anteriorly with a less pronounced anterior angulation. The nodes of the anterior angulation are usually smaller and usually arcuate with the ventral end tapering along the growth lines toward the beaks. None of the Jacksonville specimens is as well preserved as Dayville area specimens from USGS loc. 10117 and 26268, and the sculpture of the corcelet near the beaks and of the escutcheon of the Jacksonville specimens is obscured by tenaciously attached sand grains and/or leaching of the shell. It may be this difference of preservation that causes Jacksonville area specimens to seem less strongly sculptured than the Dayville specimens. The best Jacksonville area specimens are from UCLA loc. 6306 Bellinger Hill. I am indebted to the Hueners, especially Mr. Albert Hueners of Crater View Ranch for the gift of specimens culled by his grandson from rock removed to decrease the grade of Bellinger Lane over Bellinger Hill. Present among these were specimens of *"Rutitrigonia" jacksonensis* (Packard). Additional collecting from Bellinger Hill has produced *"Cucullaea" truncata* Gabb and *Meekia exotica* Saul and Popenoe in association with *Yaadia.* A similar fauna occurs on the ridge above the Jacksonville sanitary landfill (UCLA loc. 6315). Trigoniid fragments resembling *"Rutitrigonia" jacksonensis* are also present at USGS loc. M175 near Sites.

The original of fig. 178 (Gabb, 1864) (UCBMP 12171) was listed as type of Gabb's *Trigonia Gibboniana* by Merriam (1895; *in* Vogdes, 1896, p. 22; 1904 2nd ed., p. 41); Gabb proposed *T. leana* as a new name for specimens he earlier called *T. Gibboniana* Lea, and this listing by Merriam can be construed as the earliest designation of the lectotype of *T. leana.* Both Packard (1921, p. 21) and Anderson (1958, p. 113) considered Gabb's larger figure (pl. 31, fig. 262) to represent the type. Packard stated that it was from Martinez, Contra Costa County, California, but Anderson indicated Jacksonville, Oregon. Stewart (1930, p. 92) was unable to find this specimen, but he also suggested that it was from Jacksonville, Oregon, the second locality given by Gabb. It is apparent from their texts that neither Packard nor Anderson saw the

Jacksonville specimen. Three specimens labeled "*Trigonia gibboniana* Lea Jacksonville" in a handwriting that appears to be Gabb's have been found in the Museum of Paleontology, University of California, Berkeley. In the recent past they have been given UCB loc. A-6458, Siskiyou Mountains. One specimen is a shell fragment of a right valve about 58 mm. long that includes part of the beak and shows eroded noded flank ribs. The second specimen (UCBMP 14499) is a rock mold that shows the posterior corcelet margin, most of the corcelet, part of the marginal angulation, and some noded flank ribs. The third specimen (UCBMP 14498) is nearly complete but eroded especially on the beak and corcelet. It is just slightly smaller (about 3 mm.) than Gabb's fig. 262. Some of the cracks indicated in fig. 262 match those on this specimen; others do not. This specimen does not show the two curious rows of pustules on the corcelet aligned along growth lines of figure 262; but the rock mold does have such pustules, or ones that could serve as models for such pustules. Actually the rock mold indicates a small node along the median groove and similar nodes on the escutcheonal angulation and escutcheon. It is the escutcheonal nodes that Gabb appears to have transposed, placing them too far from the shell margin in a position not found in any *Yaadia*. I suggest, despite the apparent attention to cracks of fig. 262, that it was a synthetograph based upon UCBMP 14498 and 14499; both here considered to be paralectotypes.

The type locality, "Martinez" provides no age information as Gabb included material of Albian through Eocene age from various localities for several miles about under this designation. The beds south of Dayville have been recognized as Cenomanian by D. L. Jones (1960a, p. 152); UCLA loc. 3763 is above the occurrence of *Turrilites dilleri* Murphy and Rodda (1960, p. 837) in the Bald Hills Formation and thus of Cenomanian or younger? age; from CAS loc. 29580 are a specimen of *Nelltia roddana* Saul and a *Calva* sp. similar to those at Andy Bernard, Grant Co., Oregon, also of Cenomian age (Popenoe, et al., 1960, p. 1531). Bellinger Hill (UCLA loc. 6306) has yielded specimens of *Cucullaea truncata* Gabb and *Meekia exotica* Saul and Popenoe. *Cucullaea truncata* Gabb has been found in both late Albian and Cenomanian strata (Murphy and Rodda, 1960, p. 839; Murphy, 1969, p. 22), but most specimens are from reworked Cenomanian clasts in Cenomanian strata (Murphy, 1977, written notation); *Meekia exotica* was originally described from reworked late Albian clasts but has subsequently been found associated with *Anthonya cultriformis* Gabb (D. L. Jones, 1973, conversation). As Albian has not been found in the Jacksonville vicinity (Jones, 1960a, p. 154), the Bellinger Hill *Yaadia leana* are probably of Cenomanian age. Turonian ammonites are also found in Jackson Co., Oregon (Matsumoto, 1960, p. 2; Jones, 1960a, p. 154). CAS loc. 445-B, near Griffin Creek has *Y. leana*, "*Rutitrigonia*" *jacksonensis* (Packard) and *Litschkovitrigonia fitchi* (Packard). *L. fitchi* has been found in mid Turonian strata in California, and this is the only collection that associates it with *Y. leana* and "*R.*" *jacksonensis*. The beak fragment from USGS loc. M6003 occurs with *Nelltia salsa* Saul suggesting an Early Turonian age. The broken and peeled *Yaadia* from UCB loc. 2604, Elder Creek, are probably from Turonian age strata, but they are from blocks in conglomerate beds and may be reworked. *Y. leana* is at present considered to range from Cenomanian to early Turonian; but collections more carefully placed as to stratigraphic position may clarify the range of variability making possible the differentiating of Early Turonian specimens from Cenomanian ones. The holotype of *Y. leana* has relatively strong sculpture on its corcelet and is probably of Cenomanian age. Thus, although *Y. leana* may be of Turonian (early at least) age as stated by Anderson (1958, p. 109, 113) and Nakano (1960, p. 266) it is typically of Cenomanian age. *"R." jacksonensis* (Packard) which has also been said to be of Turonian age (Anderson, 1958, p. 109; Nakano, 1960, p. 265; 1963, p. 526) occurs with *Y. leana* (Gabb) at Bellinger Hill and is thus also in part of Cenomanian age (Popenoe, et al., 1960, p. 1532).

Anderson based his species *T. wheelerensis* on the specimen figured by Packard (1921, pl. 7, fig. 1, as *T. leana*) from Rock Creek, Wheeler Co., Oregon. As noted by Packard it is somewhat more strongly sculptured than specimens from the Jacksonville area and in this resembles other specimens of *Y. leana* from south of Dayville, Grant Co., Oregon. Anderson (1958, p. 109) lists *T. wheelerensis* as being of Senonian age but does not substantiate or discuss this dating. Popenoe, et al. (1960, p. 1531 and chart 10e, column 54), and Jones (1960a, p. 152) indicate that the Cretaceous beds in the Antone area which includes Rock Creek and Spanish Gulch are of Cenomanian age.

The specimen referred by Packard (1921, p. 18) to *californiana* [Calif. Acad. Sci. "12784"] and later designated holotype of *T. colusaensis* by Anderson (1958, p. 110, pl. 1, fig. 6) resembles *Y. leana* in its blocky shape. Its battered state has left only one other clue to its identity: a patch of shell remains on the anteroventral corner of the right valve, and on this the ribs are nearly straight resembling those of *Y. leana*, but they are wider than the ribs of *Y. leana* from Oregon. The specimen is also larger than any I have from Jackson Co., Oregon.

The catalogue number, 12784, given by Packard is not that of the California Academy of Sciences (=10632), but that of the California State Mining Bureau, and according to its Catalogue of the State Museum of California (1899, p. 72) CSMB 12784, *"Trigonia tryoniana* Gabb" came from "Peterson ranch, Sites, Colusa Co., California." No collector is named, but the date given is January 16, 1892. Time of deposition of the California State Mining Bureau fossil collections at the California Academy of Sciences is obscure (Rodda, 1975, conversation), but Packard (1921, p. 18) borrowed this specimen from the California Academy of Sciences. Anderson (1958, p. 110) later assigned this specimen to CAS loc. 1291 which is the type locality for *Turritella petersoni* Merriam (1941, p. 64). CAS loc. 1291 is 4 miles north of Sites, 1 mile northeast of Continental Oil Co. well, Colusa Co., California, collector, F. M. Anderson (no date given); but as the wells of the Continental Oil Company on Peterson's Ranch were drilled 1925-1926 (Stalder, 1940, p. 78), it is improbable that the *"Trigonia tryoniana* Gabb" deposited in the State Museum was collected with, at the same time as, or by the same collector as *Turritella petersoni*. Anderson assigned two other State Mining Bureau specimens to CAS loc. 1291: *"Desmoceras" colusaense* Anderson and *Turrilites petersoni* Anderson, and he (1958, p. 111) directly associates the holotype of *Trigonia colusaensis* with *Parapuzosia colusaensis* (Anderson) [*Desmoceras*]. *Pachydesmoceras colusaensis* (Anderson) [*Desmoceras*] has been collected from below the base of the Venado Formation in submarine slump deposits (Brown and Rich, 1960, p. B319) and stated by Jones (*in* Brown and Rich,

1960) to be typical of the Albian stage. Both *Mesopuzosia colusaense* (Anderson) [*Desmoceras*] and *Pseudhelicoceras petersoni* (Anderson) [*Turrilites*] were found in Late Albian deposits on the North Fork of Cottonwood Creek, Shasta Co., California (Murphy and Rodda, 1960). *Turritella petersoni* Merriam, on the other hand, has been found in the Cenomanian *Turrilites dilleri* zone (Rodda, 1959, pl. 19).

The type of specimens of *"Desmoceras" colusaense* and *"Turrilites" petersoni* from Peterson's Ranch, Colusa Co., California, have a similar very dark gray mudstone matrix, but *Trigonia colusaensis* and *Turritella petersoni* share a speckled, greenish-gray sandstone matrix. Very little of this matrix remains on the trigoniid, but that little contains fragments of *Turritella*. *"Trigonia" colusaensis* should, therefore, be associated with the horizon of *Turritella petersoni*—Cenomanian—and not that of *"Desmoceras" colusaense* and *"Turrilites" petersoni*—Late Albian. A Cenomanian age fauna which includes *Turritella petersoni* Merriam, *Cucullaea truncata* Gabb, and *Apiotrigonia condoni* (Packard) occurs at USGS loc. M176 about 75-100 feet below the base of the Venado Formation. Turonian fossils have been collected from this same vicinity (Brown and Rich, 1960, p. B319 and text fig. 149.1) from the Venado Formation. USGS loc. M175, which has a typical Turonian assemblage of bivalves and gastropods (Popenoe, 1975, conversation) including *Cucullaea gravida* Gabb, is said to be from below the base of the Venado, and may be from Venado talus. Included in this assemblage are two beak fragments and a medium-sized external mold of *Yaadia leana* (Gabb) and fragments of *"Rutitrigonia" jacksonensis* (Packard). The holotype of *"Trigonia" colusaensis* is referred to *Y. leana* on the basis of its outline and anterior truncation and considered on the basis of the *Turritella* fragments to be probably of Cenomanian age.

The specimens from CAS loc. 1294 Sand Flat, Redding area, Shasta Co., California, which Anderson identified as *"Trigonia" leana* are *Litschkovitrigonia? fitchi* (Packard).

Y. leana is readily distinguished from *Y. whiteavesi* in being less strongly truncate anteriorly with the anterior row of nodes less strongly developed, having a less strongly sculptured corcelet, and being of more equant shape. It differs from *Y.* cf. *Y. californiana* in being more truncate anteriorly, having a discernible differentiated row of nodes along the anterior angulation, and being of shorter, blockier shape.

Yaadia cf. *Y. californiana* (Packard)
(Pl. 5, figs. 5-8; pl. 6, figs. 1-2; pl. 11, fig. 4; text fig. 14, table 6)

Trigonia californiana Packard, 1921, p. 17, pl. 2, fig. 2; Stewart, 1930, p. 95; Anderson, 1958, p. 112.

Description.—Dorsal margin concavely curved. Anterior end convex with low rounded anterior angulation. Beak near anterior end. Zone of inflation along marginal angulation. Anterior angulation without distinct row of nodes, but may have very large nodes near the anterior margin. Anterior slope crossed by riblets usually to valve margin and roughened by growth lines and wrinkles. Oblique flank ribs slightly wider than interspaces, arcuate, covering the flank to the posterior end in medium-sized specimens; ribs fading near ventroposterior margin in large specimens. Corner rib with about 8 nodes. Nodes elongate along ribs with tendency to coalesce along ribs, very uneven in size. Marginal angulation nodes half as large as flank nodes, dying out about 50 mm. from beak. Nodes of outer area smaller than marginal angulation nodes, bordering the median groove, dying out 30-40 mm. from beak. Inner area crossed by fine riblets near beak which shorten at about 15 mm. from beak into nodes along the escutcheonal angulation. Escutcheon narrow, crossed by faint riblets near beak usually breaking up into scattered nodes posteriorly.

Holotype.—Lost (L. R. Kittleman, Curator of Geology, University of Oregon, 1972, letter), but there is a plaster cast of it: CAS cat. no. 568.

Hypotypes.—UCLA cat. no. 38600 from UCLA loc. 4250, 38601-38603 from UCLA loc. 6138, 48649 [or UCB cat no. 14500] from UCB loc. A-7159.

Dimensions.—See table 6.

Type locality.—UO loc. 90, Ager, Siskiyou Co., California.

Age.—Late Turonian.

TABLE 6

Measurements (in mm.) of specimens of *Yaadia* cf. *Y. californiana* (Packard)

	loc.	H	L	T	C	H/L	T/H	C/H	Remarks
38603	6138	27	33	8	15	.82	.3	.56	posterior end broken
38602	6138	33	38	11	20	.87	.33	.6	
38601	6138	75	95	17	41	.79	.23	.55	beak eroded; slightly crushed
568	90	75	109	18	44	.68	.24	.59	plaster cast of holotype; shelless
38600	4250	83	116	18?	48	.72	.22	.57	shell only along dorsal margin
48649	A-7159	83	118?	26	48?	.7	.31	.58	plaster cast, anterior edge missing
averages of ratios						.76	.27	.58	
averages of ratios H = 50 mm. +						.72	.25	.57	

H = height in sagittal plane perpendicular to ligament. L = length in sagittal plane parallel to ligament. T = thickness of one valve. C = length posterior margin of corcelet.

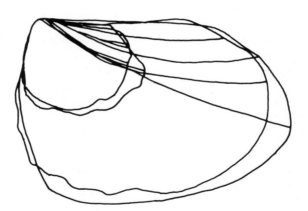

Fig. 14. Diagrammatic outlines of *Yaadia* cf. *Y. californiana* (Packard) superposed with beak and anterior margins matching as much as possible. UCLA cat. nos. 38600-38603. All drawn as left valves, margins restored, and larger specimens reduced about one third.

Remarks.—As is evident from Packard's figure (1921, pl. 2, fig. 2) and the California Academy of Sciences plastercast, the holotype lacked the outer shell layers and no ribbing is preserved. The only indication of the ribbing is low undulations on the inner shell layers remaining on the flank. Posteriorly the acute bend of the growth lines probably indicates the marginal angulation. Near the posterodorsal margin there is a faint sulcus that might indicate the position of the median groove or might be the way the shell broke. A low curved ridge on the anterior end might have been the pallial line. If it does record the pallial line, it is dissimilar to that on UCLA cat. no. 38600 which is clearly angulate at the anteroventral corner. The total locality information is quoted above. Ager is about 6 miles east of UCLA locs. 4250 and 6138 Hagerdorn-Young Ranch localities from which *Yaadia* have been collected. These four specimens of *Yaadia* were associated with a Turonian fauna, probably very late Turonian as Coniacian ammonites have been found nearby (W. P. Popenoe, 1973, conversation). Megafossils of Campanian age have also been identified from the Hornbrook Formation of that vicinity (Peck, Imlay, and Popenoe, 1956, p. 1978), but I have not seen any *Yaadia*. If the mark on the anterior end of the holotype of *Y. californiana* is the pallial line, its curve is similar to that on specimens of Late Campanian age here named *Y. robusta*. Unfortunately as the type specimen is lost no comparison can be made of rock type; the peeled cast of "*T.*" *californiana* is very suggestive of *Y. robusta,* but in the absence of definitely Campanian *Yaadia* from the Hornbrook Formation, Packard's name "*T.*" *californiana* is applied to the specimens from Hagerdorn-Young Ranch.

Anderson (1958, p. 109) lists "*T.*" *californiana* as Turonian but on page 112 states that the holotype of "*T.*" *californiana* was from an upper Senonian horizon. He provides no evidence for this assignment, an assignment made more curious by the occurrence of *Metaplacenticeras* in the vicinity of Ager, as Anderson (1958, p. 254) considered *Metaplacenticeras* indicative of Coniacian or early Senonian rather than late Senonian age. Anderson may have been influenced by Packard's (1921, p. 7) deriving "*T.*" *californiana* from "*T.*" *tryoniana* by reduction of the costae. However, it seems likely that this was a *lapsus* on Packard's part, an inadvertent switch of the position of *tryoniana* and *californiana* in the sentence, since in discussing "*T.*" *californiana* more fully he (1921, p. 18) gives the age as "Chico, and possibly Horse-town Cretaceous" and states that "*T.*" *californiana* has much more prominent ribs than "*T.*" *tryoniana* whose horizon he gives as "Upper Cretaceous, Chico group."

Yaadia cf. *Y. californiana* therefore is used for specimens of Late Turonian age. These differ from *Y. leana* in having no anterior row of nodes, more arcuate flank ribs, a slight curvature to the dorsal margin, and nodes rather than riblets along most of the escutcheonal angulation. A specimen collected and retained by Marion Ricks of Vallejo, California, from about 1 1/2 miles NW of Benicia (UCB loc. A-7159), Solano Co., California, was fortunately recorded in plaster (UCBMP cat. no. 14500) by J. H. Peck. With the exception of the anterior margin the specimen was well preserved. The low elongate shape of the valves, suggests either *Y.* cf. *Y. californiana* or *Y. robusta*. The curve of the oblique flank ribs and the nearly smooth escutcheon ornamented with a scattering of small isolated nodes indicate *Y.* cf. *Y. californiana*. The nodes on the flank ribs are not as irregular as on the specimens from northern California and in this resemble *Y. leana,* but the rib width is like that of *Y.* cf. *Y. californiana* rather than *Y. leana*.

Y. cf. *Y. californiana* is very similar in inflation, width of the oblique flank ribs and irregularity of nodes on the ribs to *Y. robusta.* These species are most easily separated by the straighter anterior margin and greater angularity of the anteroventral margin of *Y.* cf. *Y. californiana.* Specimens having the anterior end damaged are very difficult to identify, but *Y.* cf. *Y. californiana* also has slightly larger nodes on the flank ribs and the zone of inflation along the marginal angulation rather than along the second or third node of the flank ribs.

The specimen from CAS loc. 27830, said by Anderson (1958, p. 113) to be a "related species" is probably *Y. branti.* *Y. branti* is more inflated than *Y.* cf. *Y. californiana* and has a smooth inframarginal band. *Y. pinea* is also more inflated than *Y.* cf. *Y. californiana* and the nodes on the flank ribs typically run together along the growth lines rather than along the ribs.

Yaadia pinea Saul

(Pl. 6, fig. 3-6; pl. 7, figs. 1-2; text fig. 15, table 7)

Description.—Dorsal margin concavely curved. Anterior end steep but angulation rounded. Beaks near anterior end. Zone of inflation along first or second node of oblique flank ribs. Anterior angulation without distinct row of nodes, but some specimens have a few larger nodes along the angulation. Anterior slope roughened by growth lines usually without ribbing or nodes near the valve margin. Anterior angulation nodes usually aligned with oblique flank ribs and not separated by interspace. Oblique flank ribs slightly wider than interspaces, covering the flank to the posterior end but becoming lower near ventroposterior margin. Corner rib with about 9 nodes. Mid flank nodes usually elongated and coalesced along ribs; marginward nodes strongly elongated and coalesced along growth lines to form concentric ribs near anteroventral and posteroventral margins. Nodes and ribbing notably irregular. Marginal angulation nodes small, dying out about 30-40 mm. from beak. Nodes of outer area small, bordering obscure median groove, dying out 20-30 mm. from beak. Inner area crossed by riblets near beak, as area widens riblets remain about 4 mm. long, lie along escutcheon boundary, die out about 20-30 mm. from beak. Escutcheon not delineated by angulation, nearly smooth, narrow.

Holotype.—UCLA cat. no. 38604, UCLA loc. 6304.

Paratypes.—UCLA cat. nos. 38605-10, UCLA loc. 6304.

Dimensions.—See table 7.

Type locality.—UCLA loc. 6304, Pine Timber Gulch, Shasta Co., California.

Age.—Coniacian.

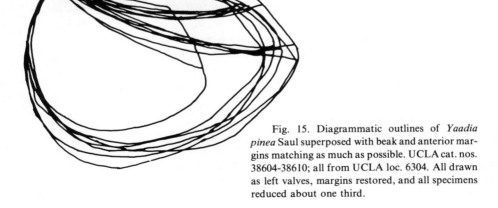

Fig. 15. Diagrammatic outlines of *Yaadia pinea* Saul superposed with beak and anterior margins matching as much as possible. UCLA cat. nos. 38604-38610; all from UCLA loc. 6304. All drawn as left valves, margins restored, and all specimens reduced about one third.

TABLE 7

Measurements (in mm.) of specimens of *Yaadia pinea* Saul

	loc.	H	L	T	C	H/L	T/H	C/H	Remarks
38607	6304	46?	72	13	30	.64	.28	.65	broken anteroventrally
38608	6304	50	56	14	31	.89	.28	.62	broken ventrally
38606	6304	54	84	21	30	.64	.39	.56	shell fractured
									H & C shortened
38605	6304	57	82	18	27?	.7	.32	.47	dorsal margin broken
38610	6304	?	88	19	30	?	?	?	broken
38604	6304	59	93	20	37	.63	.34	.63	holotype, beak
									eroded
38609	6304	60	87	20	31	.69	.33	.51	beak eroded
averages of ratios						.7	.32	.57	
averages of ratios H = 50 mm. +						.71	.33	.55	

H = height in sagittal plane perpendicular to ligament. L = length in sagittal plane parallel to ligament.
T = thickness of one valve. C = length posterior margin of corcelet.

Remarks.—The species is based upon 13 specimens from UCLA loc. 6304, Pine Timber Gulch, Shasta Co., California. Two other specimens of Coniacian age have been found in the Redding area at CIT loc. 1232 and UCLA loc. 6010, but both are nearly shellless and do not add to the definition of the species. Additional scrappy specimens occur in a Panoche conglomerate on Sagaser Hill, UCLA loc. 6312, Garza Peak quad., Kings Co., California. Unfortunately the beak area is not well preserved in any available specimen. UCLA cat. no. 38604 is chosen as holotype because of its relatively complete outline. Sculpture is better preserved on UCLA cat. nos. 38606 and 38607. There is considerable variability in shape and sculpture of this species. As all the specimens with shell are from the same locality and from a very limited stratigraphic interval, these differences cannot be ascribed to geographic or time separation. This variability makes defining *Y. pinea* difficult.

It differs from *Y.* cf. *Y. californiana* (Packard) mainly in the more curved dorsal margin, in being more inflated anteriorly, in having the anterior face usually un-ribbed, and the strong tendency for the nodes of the flank ribs to coalesce along the growth lines to form concentric ribs. Only in *Y. hemphilli* are similar ribs formed to a similar degree. *Y. pinea* lacks the intricate near beak ornamentation of *Y. hemphilli*. *Y. pinea* resembles *Y. branti* in shape but in *Y. pinea* the flank ribs start at the marginal angulation and there is not the distinctive unribbed wedge anterior to the marginal angulation of *Y. branti*.

The species is named for its occurrence in Pine Timber Gulch.

Yaadia branti Saul
(Pl. 7, figs. 3-10; pl. 8, fig. 1; pl. 11, fig. 5; text fig. 16; table 8)

Description.—Dorsal margin concavely curved. Anterior end flatly truncated, high, usually with well-developed angulation. Beaks near anterior end. Zone of inflation between third and fifth node of flank ribs. Anterior angulation without differentiated row of nodes. Anterior slope usually roughened by growth lines, some specimens with narrow riblets and/or nodes. Oblique flank ribs nearly twice as wide as interspaces, not covering the flank to the posterior end. Smooth wedge-shaped inframarginal band bordering the marginal angulation. Corner rib with about 9 nodes. Flank nodes elongated and coalesced along ribs; marginward nodes elongated and coalesced along growth lines. Marginal angulation nodes slightly smaller than neighboring flank rib nodes, disappearing 30-40 mm. from beak. Nodes of outer area small, bordering well-marked median groove, disappearing 30-40 mm. from beak. Inner area crossed by riblets near beak; about 20 mm. from beak riblets become nodes which lie on the slight escutcheonal angulation. Escutcheon usually smooth, narrow.

Holotype.—UCLA cat. no. 38613.

Paratypes.—UCLA cat. nos. 38616 from UCLA loc. 3621, 38618-21 from CIT loc. 1016, 38623 from UCLA loc. 3623, 38625 from UCLA loc. 3625.

Dimensions.—See table 8.

Type locality.—UCLA loc 3619, Chico Creek, Paradise quad., Butte Co., California.

Age.—Santonian.

Fig. 16. Diagrammatic outlines of *Yaadia branti* Saul superposed with beak and anterior margins matching as much as possible. UCLA cat. nos. 38613, 38616, 38618-38620, 38623. All drawn as left valves, margins restored, and larger specimens reduced about one third.

Remarks.—This species is based upon more than 20 specimens from 7 localities in the lower part of the Chico Formation along Chico Creek, Butte Co., California. The seven localities are within a stratigraphic interval of 500 feet in beds dated as Santonian by Matsumoto (1960, p. 18). Two beak fragments from CAS loc. 28102, Basin Hollow, and a broken, eroded double valved specimen from CAS 27830, 4.9 miles east of Millville, Shasta Co., California, are probably this species. The specimens from Chico Creek referred to *"T." hemphilli* by Anderson (1958, p. 115) are undoubtedly *Y. branti*.

The smooth inframarginal band makes *Y. branti* one of the easier species of *Yaadia* to distinguish. It is broader beaked than *Y. pinea;* it has a steeper, more angulate anterior face and a less inflated posterior than *Y. tryoniana*.

The species is named for Richard Brant Saul who helped collect many of the specimens.

TABLE 8

Measurements (in mm.) of specimens of *Yaadia branti* Saul

	loc.	H	L	T	C	H/L	T/H	C/H	Remarks
38621	1016	14	15+	6	7	.93	.43	.5	posterior broken
38619	1016	17	22	6	11	.77	.35	.65	
38620	1016	19	24	6	10	.79	.32	.53	
38625	3625	61	90	17	33	.68	.28	.54	
38623	3623	63	95	20	37	.66	.32	.59	
38613	3619	66	94	25	39	.7	.38	.59	holotype
38616	3621	67	92	24	37	.73	.36	.55	
38618	1016	70	97	27	35	.72	.39	.5	
averages of ratios						.75	.35	.56	
averages of ratios H = 50 mm. +						.7	.34	.55	

H = height in sagittal plane perpendicular to ligament. L = length in sagittal plane parallel to ligament. T = thickness of one valve. C = length posterior margin of corcelet.

Yaadia tryoniana (Gabb)

(Pl. 8, figs. 2-7; pl. 9, fig. 1; text fig. 17, table 9)

Trigonia tryoniana Gabb, 1864, p. 188, pl. 25, fig. 176; Whiteaves, 1879, p. 161, pl. 18, fig. 7; Packard, 1921, p. 19, pl. 4, fig. 4; Stewart, 1930, p. 94; Anderson, 1958, p. 114.

Description.—Dorsal margin concavely curved. Anterior end slightly truncated, rounded. Beak near anterior end. Zone of inflation along second or third node of flank ribs. Anterior angulation without differenciated row of nodes. Anterior slope roughened by growth lines. Oblique flank ribs nearly twice as wide as interspaces, becoming very low near ventro-posterior margin. Corner rib with about 13 nodes. Flank nodes elongated and coalescing along growth lines near the ventral margin. Marginal angulation nodes as large as or larger than adjacent flank rib nodes, present up to 55 mm. from beak. Nodes of outer area half as large as marginal angulation nodes, elongate in growth line direction, present up to 45 mm. from beak, lying along clearly marked median groove. Inner area crossed by riblets near beak, present up to 45 mm. from beak. Riblets near beak posteriorly directed at angle of about 30° to dorsal valve margin, becoming normal to valve margin about 25 mm. from beak, more posterior riblets roughly parallel to growth lines. Juvenile ribbing of corcelet forming obvious chevrons up to 20 mm. from beak. Escutcheon delimited by rounded angulation, a little depressed, usually smooth.

Holotype.—UCBMP cat. no. 11955.

Hypotypes.—CGS cat. no 5834; UCLA cat. nos. 38628 from UCLA loc. 4082, 38630-38632 from CIT loc. 1169; USNM cat. no. 241680 from USGS loc. 405; CAS cat. no. 57981 from CAS loc. 2365.

Dimensions.—See table 9.

Type locality.—Tuscan Springs, Little Salt Creek, Tehama Co., California.

Age.—Early Campanian, *Submortoniceras chicoense* zone.

Remarks.—Although more than a hundred years have passed since its original description, there is as yet no good specimen of *Y. tryoniana*. The holotype has the outermost layer of shell removed thus leading to the erroneous impression that *Y. tryoniana* is less strongly ribbed than other related species (Packard, 1921, p. 18). This peeling of the shell has also caused the radiating grooves mentioned by Gabb (1864, p. 188) and Packard (1921, p. 19). The above description is based upon the holotype, a beak fragment (UCLA cat. no. 32628) from the type locality, Whiteaves' specimen from Northwest Bay, Vancouver Island, B.C., and Turner's (1894, p. 460, USGS loc. 405) specimen from Pentz Ranch, Butte Co., California. Berkeley has two scraps

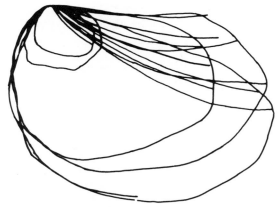

Fig. 17. Diagrammatic outlines of *Yaadia try-*
oniana Gabb superposed with beak and anterior
margins matching as much as possible. CAS cat.
no. 57981; UCLA cat. nos. 28958, 32199, 38628,
38630; USNM cat. no. 241680. All drawn as left
valves, margins restored except for USNM 241680,
and larger speicmens reduced about one third.

TABLE 9

Measurements (in mm.) of specimens of *Yaadia tryoniana* (Gabb)

	loc.	H	L	T	C	H/L	T/H	C/H	Remarks
57981	2365	14	18	4	10	.78	.29	.71	
38628	4082	21	22?	6	11	.95	.29	.52	posterior broken?
28958		55	76?	14	32?	.72	.25	.58	plaster cast of skinned holotype
32199		58	90	17	36	.64	.29	.62	plaster cast of Whiteaves' hypotype
38630	1169	71	108	27	35	.66	.38	.49	
241680	405	80	?	30	?	?	.38	?	posterior broken, shell missing ventrally
38631	1169	85	116+	26	39?	.73	.31	.46	dorsoposterior end broken, shell missing ventrally
averages of ratios						.75	.31	.56	
averages of ratios H = 50 mm. +						.68	.32	.53	

H = height in sagittal plane perpendicular to ligament. L = length in sagittal plane parallel to ligament.
T = thickness of one valve. C = length posterior margin of corcelet.

(Calif. Geol. Surv. 185) labeled in Gabb's handwriting "*Trigonia Hanetiana* d'Orb.,
Tuscan Springs." The label has been corrected (hand unknown) to "*tryoniana* G."
which they may well be; although noded ribs are better preserved on these scraps than
on the holotype, they retain no characteristics helpful in distinguishing the species.
California Academy of Sciences has a large scrap of a *Yaadia* from CAS loc. 31334,

Pigeon Point Formation, just north of Bolsa Point, San Mateo Co., California. Very little shell remains on this specimen, but enough is retained on the anterior margin to indicate that it is not *Y. hemphilli* (Anderson). I have seen only a plaster cast of Whiteaves' specimen (1879, pl. 18, fig. 7) CGS 5834 from Northwest Bay, Vancouver Island. The beak ribbing forms chevrons like those on UCLA 38628 from Tuscan Springs, Tehama Co., California. Unfortunately the shell appears to have been peeled from the inner area and escutcheon removing the riblets that are helpful in distinguishing *Y. tryoniana.* The anterior face is somewhat steepened and apparently not ribbed to the margin and therefore suggestive of *Y. tryoniana.* The broadly rounded inflation of the shell is also suggestive of *Y. tryoniana.* Jeletzky (*in* Muller and Jeletzky, 1970, p. 42, fig. 7, sec. 7) considers the Northwest Bay sandstone facies of the Haslam Formation to carry the *Inoceramus schmidti* fauna which contains *Canadoceras yokoyami* and *C. multisulcatum.* If Whiteaves' specimen is correctly assigned to *Y. tryoniana,* the beds at Northwest Bay, Vancouver Island, B.C., should correlate with the upper Chico Formation of Early Campanian age containing *Submortoniceras chicoense, Canadoceras yokoyami,* and *C. multisulcatum.* Certainly Anderson (1958, p. 114) errs in assigning this species an age not younger than Santonian, as it occurs with *Submortoniceras chicoense* at Pentz, Butte Co., California. Four imperfect specimens from CIT loc. 1169, Baker Canyon, Santa Ana Mts., California, in the "*Turritella chicoensis perrini* division" of Popenoe (1942), resemble *Y. tryoniana,* but do not show obvious chevron ribbing of the juvenile area. The nodes of the outer area are not elongate and the ribbing of the inner area is stronger. Scraps of *Yaadia* also occur at CAS loc. 2365, northwest of Big Tar Canyon, Kings Co., California; most are undeterminable, but one juvenile (CAS cat. no. 57981) resembles the beak area of specimens from CIT 1169.

    Yaadia tryoniana differs from *Y. pinea* in being more rounded anteriorly. It differs from *Y. robusta* in having a steeper and smoother anterior face. The inner area riblets of *Y. robusta* are shorter on the inner area but more often continue across the escutcheon, and they are posteriorly directed. *Y. tryoniana* differs from *Y. hemphilli* in its steeper smoother anterior face and its usually smooth escutcheon.

Yaadia robusta Saul
(Pl. 9, figs. 2-4, 6; text fig. 18, table 10.)

Description.—Dorsal margin concavely curved. Anterior end rounded. Beaks near anterior end. Zone of inflation along second node of flank ribs. Anterior without marked angulation or distinctive row of nodes. Flank ribs extending to or nearly to the anterior margin. Oblique flank ribs about 1.5 times as wide as interspaces, becoming very low near ventroposterior margin. Corner rib with about 10 nodes. Flank nodes with tear-drop shape anteromedially, coalescing along ribs posteriorly, elongate along growth lines ventrally. Marginal angulation nodes about same size as adjacent flank rib nodes, present up to 60 mm. from beak. Nodes of outer area half as large as marginal angulation nodes, present up to 50 mm. from beak, bordering clearly marked median groove. Inner area crossed by riblets near beak, becoming nodes bordering the escutcheon, nodes twice the size of marginal angulation nodes, continuing to 70 mm. from beak. Escutcheon crossed by posteriorly directed riblets from inner area nodes to valve margin.

    Holotype.—UCLA cat. no. 38634.

    Paratypes.—UCLA cat. nos. 38756-38758 from UCLA loc. 6310.

    Hypotypes.—UCLA cat. nos. 38635 from CIT loc. 86, 38636 from UCLA loc. 6298.

    Dimensions.—See table 10.

    Type locality.—CIT loc. 1159: Dayton Canyon, Simi Hills, Los Angeles Co., California.

    Age.—Late Campanian, *Metaplacenticeras pacificum* zone.

Fig. 18. Diagrammatic outlines of *Yaadia ro-busta* Saul and *Y.* cf. *Y. robusta* Saul superposed with beak and anterior margins matching as much as possible. UCLA cat. nos. 38634, 38636, 38637, 38756. All drawn as left valves, margins restored, and all reduced about one third.

Remarks.—The above description is based on the well-preserved holotype and three incomplete paratypes from UCLA loc. 6210, mouth of Topanga Canyon, Los Angeles Co., California. The fragmental hypotype from the Pleasants Sandstone of the Santa Ana Mountains (CIT loc. 86) agrees with these four specimens in all preserved features. The hypotype from the Panoche Formation, Los Gatos Creek (UCLA loc. 6298), Fresno Co., California, differs in having less well-developed sculpture on the corcelet and escutcheon. Some sculpture may have been lost by abrasion before preservation; some sculpture was lost in grinding the hard matrix from the recrystallized shell. The sculpture of its corcelet and escutcheon resembles

TABLE 10

Measurements (in mm.) of specimens of *Yaadia robusta* Saul

	loc.	H	L	T	C	H/L	T/H	C/H	Remarks
38756	6310	63	89	26	30	.71	.41	.48	
38637	2415	64+	80+	25	36?	.8	.39	.56	ventral and posterior margins missing
38636	6298	70+	109+	22	46	.64	.31	.66	anterior margin missing
38638	2415	72	94+	26	33	.77	.36	.46	
38634	1159	78	93	16	38	.84	.2	.49	holotype
averages of ratios						.75	.33	.53	all with H = 50 mm. +

H = height in sagittal plane perpendicular to ligament. L = length in sagittal plane parallel to ligament. T = thickness of one valve. C = length posterior margin of corcelet.

that of *Y.* cf. *Y. robusta* from UCLA loc. 2415, Bee Canyon, Santa Ana Mountains, but the Los Gatos Creek specimen is less inflated.

In shape this species is most similar to *Y. californiana* from which it can be separated by the less anterior position of the beak and more rounded anterior margin, the less angulate anteroventral margin, and the more obtuse angle of the growth lines crossing the marginal angulation. The plaster cast from UCB loc. A-7159, Benicia, Solano Co., California, and the Los Gatos Creek specimen (UCLA cat. no. 38636) are nearly identical in shape and both have been deprived of their anterior margins. The Los Gatos Creek specimen is more rudely noded, but a specimen of *Y.* cf. *Y. californiana* from UCLA loc. 6138, Hornbrook Formation, Siskiyou Co., California, is also rudely noded. The Los Gatos Creek specimen has elongate inner area nodes and the escutcheonal riblets associated with them posteriorly directed; but the Benicia specimen has the inner area and escutcheonal riblets ligamentally directed, and these riblets become nodular closer to the beak. The flank ribs of *Y. robusta* are closer spaced than those of *Y.* cf. *Y. californiana* and the zone of inflation for *Y. robusta* lies along the second node of the flank ribs rather than along the marginal angulation as in *Y.* cf. *Y. californiana.*

Scraps of *Yaadia* from CAS loc. 33704, Quinto Creek, Merced Co., California, have very irregularly noded, arcuate flank ribs and a low anterior profile with no anterior row of nodes and may be this species. Packard (1916, p. 147) lists *Trigonia tryoniana* Gabb from the shallow water phase of his *Turritella pescaderoensis* zone, but I was unable to find specimens from the localities he lists (1916, p. 145) as characteristic of this phase and zone or from any other of Packard's Santa Ana Mts. localities in the UCB collections. UCB loc. A-6446 which is from the H. W. Fairbanks collection contains three fragmental specimens. Possibly these are the specimens for Packard's reference. The block of rock containing two of the specimens also has *Meekia (Mygallia) daileyi* Saul and Popenoe and *Cymbophora popenoei* Saul. Both Packard and Popenoe found *Meekia daileyi* Saul and Popenoe (1962, p. 312) only in the *Metaplacenticeras pacificum* zone of the Santa Ana Mts. These poorly preserved *Yaadia* are probably *Y. robusta.*

Y. robusta differs from *Y. tryoniana* in its more sloping anterior upon which the flank ribs extend to or nearly to the valve margin. It differs from *Y. hemphilli* in having posteriorly directed, less strongly developed sculpture on the escutcheon, and in having larger nodes along the anterior margin.

Yaadia cf. *Y. robusta* Saul

(Pl. 9, fig. 5; pl. 10, fig. 1; text fig. 18, table 10.)

Description.—As for *Y. robusta* with the following exceptions: Anterior end a little truncated with slight angulation. Flank ribs dying out near the anterior valve margin. Corner rib with about 9 nodes. Flank nodes not tear-drop shaped. Marginal angulation nodes smaller than adjacent flank rib nodes, present to about 35 mm. from beak. Nodes of outer area present up to about 45 mm. from beak. Inner area sculpture dying out about 50 mm. from beak. Escutcheon delimited by angulation, nearly smooth, ornamented near beak and along flankward margin by posteriorly directed riblets associated with inner area nodes.

Hypotypes.—UCLA cat. no. 38637-38638 from UCLA loc. 2415.

Dimensions.—See table 10.

Age.—Very late Campanian (see Matsumoto, 1960, p. 154, UCLA loc. 2415) associated with *Turritella pescaderoensis* Arnold.

Remarks.—These two specimens differ from other *Y. robusta* in being more inflated and having less persistent sculpture on the corcelet and escutcheon. This greater inflation causes these two specimens to resemble *Y. hemphilli,* but their inner area riblets are directed as in *Y. robusta.* These differences from *Y. robusta* s.s. are minor and may signify only specific variability. The specimens from UCLA loc. 2415 are discussed separately here because only they show these variations and they may be geologically younger than typical *Y. robusta.*

Yaadia hemphilli (Anderson)

(Pl. 10, figs. 2-9; pl. 11, figs. 3, 6; text figs. 8, 19; table 11.)

Trigonia leana var. *whiteavesi* Packard, 1921, p. 21, pl. 5, fig. 4 only.
Trigonia hemphilli Anderson, 1958, p. 115, pl. 52, fig. 9, 9a-9b.

Description.—Dorsal margin concavely curved. Anterior end rounded. Beak near anterior end. Zone of inflation along second or third node of flank ribs. Anterior usually without marked angulation and without differentiated row of nodes. Anterior slope with narrow extensions of flank ribs to valve margin. Oblique flank ribs a third wider than interspaces, becoming very low near ventroposterior margin. Corner rib with about 11 nodes. Flank nodes elongated and coalescing along growth lines near the ventral margin. Marginal angulation nodes nearly as large as adjacent flank rib nodes, present up to 65 mm. from beak, becoming elongate in the growth line direction posteriorly. Outer area ribbed near beak. Ribs shortening to nodes 20 mm. from beak. Nodes of outer area half as large as marginal angulation nodes, bordering well-marked median groove, present to 50 mm. from beak. Inner area crossed by riblets near beak, present up to 45 mm. from beak. Juvenile ribbing of area forms complicated chevrons. Escutcheon narrow, a little depressed, crossed by ligamentally directed riblets from inner area, riblets to valve margin. Commissure of young specimens waved by alternation of ribbing on two valves.

Holotype.—CAS cat. no. 994.

Hypotypes.—UCLA cat. nos. 38639 from UCLA loc. 4141, 38640 from UCLA loc. 4144, 38645 from CIT 590, 38650 from UCLA loc. 4122, 38651 from UCLA loc. 4124, 38658 from UCLA loc. 4132, 38659 from UCLA loc. 4138, 38661-62 from UCLA loc. 4134, 38663 from UCLA loc. 4127, 38665 from UCLA loc. 4133, 44424 from UCLA loc. 4118.

Dimensions.—See table 11.

Type locality.—CAS loc. 12123, "near Pescadero, California" (Anderson, 1958, p. 115) but CAS catalogue book says "Tertiary fossils, bivalves, Pescadero, California" (Rodda, 1972, letter).

Age.—Early Maestrichtian.

Remarks.—In addition to the holotype, the above description is based upon more than 30 specimens from the Jalama Formation, southwestern Santa Barbara County, California (Dailey and Popenoe, 1966, fig. 3). This is undoubtedly the "highly ornate variety" of *Trigonia leana* from the Santa Ynez Mountains, California, mentioned by Packard (1921, p. 7). It has not been found at any locality on Chico Creek or from the Chico Formation despite Anderson's statements (1958, p. 115). It does not occur at CAS loc. 27838 as listed by Anderson (1958, p. 45). The specimens from Chico Creek are *Y. branti* and differ from *Y. hemphilli* in having the smooth inframarginal band, a more truncate anterior end, and less sculpture on the area and escutcheon.

Packard (1921) on p. 22 gives the locality of the specimen he figures on plate 5, figure 4 (= holotype of *T. hemphilli* Anderson) as Pigeon Point, San Mateo Co., California, but I have been unable to find this species in the vicinity of Pigeon Point. A fragment from CAS loc. 31334 about a mile north of Pigeon Point is smooth along the anterior margin and therefore questionably identified as *Y. tryoniana.* The presence of *Meekia (Mygallia) bella* Saul and Popenoe (1962, p. 310) at this locality suggests an Early Campanian age for these beds of the Pigeon Point Formation. In the explanation to plate 5, Packard gives the locality as Pescadero as does the

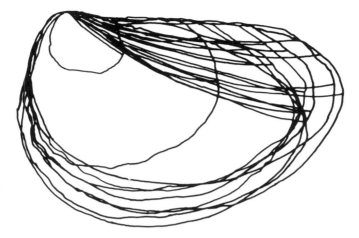

Fig. 19. Diagrammatic outlines of *Yaadia hemphilli* (Anderson) superposed with beak and anterior margins matching as much as possible. UCLA cat. nos. 38639, 38645, 38650-51, 38658-38659, 38661-38663, 38665. All drawn as left valves, margins restored, and larger specimens reduced about one third.

TABLE 11

Measurements (in mm.) of specimens of *Yaadia hemphilli* (Anderson)

	loc.	H	L	T	C	H/L	T/H	C/H	Remarks
38651	4124	17	21	6	9	.81	.35	.53	
44424	4118	34	41	10	16	.83	.29	.47	
38662	4134	51	59	14	24	.86	.26	.47	
38639	4141	62	93	20	33	.67	.32	.53	
38659	4138	64	96	23	39	.67	.35	.61	
38661	4134	66	87	23	36	.76	.35	.54	
38663	4127	67	100	23	36	.67	.34	.54	
48659	12123	70	?	24	?	?	.34	?	plaster cast of holotype, posterior end lopped off
38658	4132	82	110	24	38	.74	.29	.46	beak eroded, anterior broken
averages of ratios						.75	.32	.52	
averages of ratios H = 50 mm. +						.73	.32	.52	

H = height in sagittal plane perpendicular to ligament. L = length in sagittal plane parallel to ligament. T = thickness of one valve. C = length posterior margin of corcelet.

California Academy of Sciences catalogue book. The catalogue entry seems strange, however, in mistaking a trigoniid for a Tertiary bivalve. Hall, Jones, and Brooks (1959, fig. 2) map Pliocene age Purisima Formation in the immediate vicinity of Pescadero. The Pigeon Point Formation crops out along Pescadero Point; and at UCB loc. A-7601, about one mile north of Pescadero Point, ammonite fragments referable to *Didymoceras* and therefore suggestive of Maestrichtian age were found (Hall, *et al., 1959, p. 2858—fide* J. W. Durham and J. E. Peck, 1973, conversation, locality is not one mile south as given in Hall, *et al.,* but instead one mile north). No *Yaadia* has been found in this vicinity and the holotype of *Y. hemphilli* does not have the same matrix as the specimens of *Didymoceras.* Its matrix is very similar to that of specimens from the Jalama Formation, but the Pigeon Point Formation also includes sandstone beds of similar aspect. Had the provenance of the holotype been unclouded and precise, it might have improved dating in the Pigeon Point Formation. The occurrence of *Y. hemphilli* in the Jalama Formation which contains a fauna probably slightly younger than the zone of *Metaplacenticeras* (Dailey and Popenoe, 1966, p. 1) indicates that it is Early Maestrichtian in age, not "lower Campanian" as stated by Anderson (1958, p. 115).

   Y. hemphilli differs from *Y. robusta* in having longer, ligamentally directed inner area riblets crossing the escutcheon, more arcuate flank riblets, and a greater tendency for the nodes of the flank ribs to coalesce along growth lines near the ventral margin. *Y. hemphilli* differs from *Y. tryoniana* in its more rounded anterior with flank ribs extending to the shell margin.

Quadratotrigonia Dietrich, 1933
Type-Species by original designation:
Trigonia nodosa J. de C. Sowerby, 1826
Quadratotrigonia sp. A
(Pl. 11, fig. 10.)

Description.—Shell large, quadrate, roundly inflated. Sculpture of anterior end neatly aligned noded ribs curving from the margin toward the marginal angulation. Flank apparently sculptured by wider ribs bearing larger nodes but degree of regularity cannot be determined because so little of the flank is preserved. Marginal angulation apparently marked by row of nodes; median carina marked by row of nodes, escutcheon marked by angulation and row of nodes. Inner and outer areas pustulate; escutcheon noded.

Hypotype.—UCLA cat. no. 38764 from UCLA loc. 3882, south flowing tributary to Clear Creek north of Reading Bar, Redding quad., Shasta Co., Calif., Budden Canyon Formation.

Dimensions.—Height 84 mm; length 89 mm. (broken); thickness 23 mm.

Age.—Late early Albian; *Brewericeras hulenense* zone.

   Remarks.—The anterior end of the hypotype is adequately preserved, but the rest of the shell surface is either eroded or missing. There is also a poorly preserved second specimen, the dorsal half of a right valve, from the same Clear Creek tributary (UCLA loc. 3899). The shell of *Q.* sp. A has a steepened anterior slope with differentiated noded ribs after the style of *Q. daedalea* Parkinson which is distinctly different from the more sloping anterior profile of *Q. mearnsi* (Stoyanow). *Q.* sp. A appears to have had smaller flank nodes than *Q. mearnsi* and larger flank nodes than *Q. taffi* (Cragin). *Q.* sp. A is obviously new, but the specimens are too poorly preserved to describe a species.

Quadratotrigonia sp. B
(Pl. 11, figs. 9, 11)

Description.—Shell probably quadrate with a rounded anterior angulation. Angulation marked by a row of nodes, from which low noded ribs extend toward the anterior margin. Flank sculptured by broad, very nodose, nearly straight ribs angling from the marginal angulation toward the valve margin; interspaces broad.

Hypotype.—UCLA cat. no. 38837 from HSU loc. 1277, Cedros Island, Baja California, Mexico; Valle Formation, upper member.

Dimensions.—Height 61 mm.; length 70 mm.; thickness 23 mm.

Age.—?Turonian.

Remarks.—Q. sp. B has a more strongly developed anterior differentiation with steeper slope than *Q.* sp. A. The flank is sculptured by broader, fewer, straighter oblique ribs. The ribs interspaces of *Q.* sp. B are wider than those of *Q. mearnsi* (Stoyanow). Unfortunately the corcelet is preserved only very near the beak where fine chevron riblets resembling those of *Q. deslongchampsi* (Munier-Chalmas) are present. A corner of the outer area is present at the valve margin. This portion is smooth. *Q.* sp. B is also apparently new; the shell that is present is well preserved, but the missing corcelet makes description inadvisable. Its Turonian age is tenuously based upon its occurrence with a *Pyrazus* n. sp. which at a nearby locality occurs with specimens identified as *Trigonarca californica* Packard.

Litschkovitrigonia Saveliev, 1958
Type-Species by original designation:
Trigonia litschkovi Mordvilko, 1953
Litschkovitrigonia? *fitchi* (Packard)
(Pl. 12, figs. 1-7; text fig. 20, table 12)

Trigonia fitchi Packard, 1921, p. 20, pl. 6, fig. 3; pl. 7, fig. 2.
Trigonia branneri Anderson, 1958, p. 112, pl. 17, fig. 5.

Description.—Outline nearly ovate with moderate beaks and posterior truncation. Maximum inflation of valves along nearly straight marginal angulation; anterior not angulate, rounded to the valve margins, corcelet roundly inflated. Sculpture of flank variable, ribs from anterior valve margin to marginal angulation sometimes oblique curved, sometimes with jogs or indistinct V; nodes small, may be coalesced along growth lines or into v patterns; concentric growth checks (?) occasionally as strong as oblique ribs. Corcelet broad, ornamented near beak, smooth posteriorly, divided by well-marked median groove into broad outer area and narrow inner area. Outer area crossed by fine riblets obliquely angled toward beaks, evanescing about 15 mm. from beak. Inner area crossed by fine riblets obliquely angled away from beak, evanescing about 20 mm. from beak. Escutcheon very broad, defined by obtuse angulation, crossed by fine riblets obliquely angled away from beak and persisting to 30-40 mm. from beak.

Holotype.—UO cat. no. 26859 (plaster cast UCLA 48656).

Paratype.—UO cat. no. 26859 (plaster cast UCLA 48657) from UO loc. 98 near Ager, Siskiyou Co., California.

Hypotypes.—LSJU cat. no. 8697 = holotype of *T. branneri* Anderson (plaster cast UCLA cat. no. 48661) from Rocky Gulch 2 1/2 miles SW of Hornbrook, Siskiyou Co., California; UCLA cat. nos. 38760-38761 from UCLA loc. 6140, 38762 from UCLA loc. 6305.

Dimensions.—See table 12.

Type locality.—UO loc. 93, Griffin Creek, Jackson Co., Oregon.

Age.—Early and Mid Turonian.

Remarks.—L. fitchi differs from all West Coast *Yaadia* in its less prominent beaks, lower more rounded anterior, sculptured by closely spaced rows of small nodes, the pitched-roof slope of its inflation away from the marginal angulation with no flat-

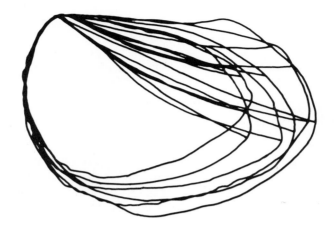

Fig. 20. Diagrammatic outlines of *Litschkovi-trigonia? fitchi* (Packard) superposed with beak and anterior margins matching as much as possible. UCLA cat. nos. 38760-38762, 48656-48657, 48661. All drawn as left valves, margins restored, and all reduced about one third.

TABLE 12

Measurements (in mm.) of specimens of *Litschkovitrigonia? fitchi* (Packard)

	loc.	H	L	T	C	H/L	T/H	C/H	Remarks
48656	93	48	65	13	?	.74	.27	?	plaster cast of holo-type UO cat. no. 26859
48657	98	50	75	17	27	.67	.34	.54	left valve of plaster cast of paratype UO cat. no. 26859
48661		65	86	17	37	.76	.26	.6	plaster cast of *Trigonia branneri* Anderson LSJU cat. no. 8697
38760	6140	57	82	22	32	.7	.38	.56	
38761	6140	50	66	16	33	.76	.32	.66	
38762	6305	60	85	21	37	.71	.35	.62	
averages of ratios						.72	.32	.6	

H = height in sagittal plane perpendicular to ligament. L = length in sagittal plane parallel to ligament. T = thickness of one valve. C = length posterior margin of corcelet.

tened curve to the flank, and its narrower more closely spaced flank ribs. The inhalant-exhalant areas are bounded internally by short but obvious ridges, the ridge separating the two areas being the strongest. Both valves appear to have had two low "lateral teeth" and sockets at the posterior end of the dorsal margin adjacent to the exhalant area. In *Yaadia* similar "teeth" are entirely posterior to the posterior adductors, but in *L. fitchi* the "teeth" are slightly more anteriorly placed being above the posterior side of the posterior adductor muscle scars.

Assignment of *"T." fitchi* to *Litschkovitrigonia* is tentative. It resembles *Korobkovitrigonia korobkovi* Saveliev in shape, width of corcelet and divisions of corcelet, but the flank ribs of *"T." fitchi* do not form well-marked chevrons near the beak. *"T." fitchi* is also similar in shape and flank ribbing to *Linotrigonia danovi* Saveliev (1958, pl. 40, figs. 1-3), but *"T." fitchi* has a more inflated corcelet the sculpture of which is clearly demarked by the marginal angulation, median groove and escutcheonal angulation. *Linotrigonia* is included in the Pterotrigoniinae (e.g., Saveliev, 1958, p. 116), but the shape and sculpture especially of the corcelet of *"T." fitchi* are more suggestive of the Myophorellinae. *"T." fitchi* is more oval than *Litchkovitrigonia ovata* (Litschkov) and has irregular ribs along the anterior margin which usually approach the valve margin at a ventrally directed angle, whereas the anterior ribbing of *L. ovata* is more regular and normal to the valve margins. The sculpture of the corcelet of *"T." fitchi* and *L. ovata* is similar.

L. fitchi has been found in the Hornbrook Formation, Siskiyou Co., California, and Jackson Co., Oregon and in Member I, Redding area, Shasta Co., California. Packard's locality data gives no indication of stratigraphic position, but Anderson (1958, p. 112) indicated the lower beds of the Turonian for *T. branneri* from Rocky Gulch. UCLA loc. 6140, Rocky Gulch, is in Member I and thus probably of Mid Turonian age (Peck, Imlay, and Popenoe, 1956). In addition to a specimen of *L. fitchi*, CAS loc. 445-B, near Griffin Creek, Jackson Co., Oregon, has *Y. leana* and *"Rutitrigonia" jacksonensis* which might indicate an age as early as Cenomanian. Anderson (1958, p. 20) recognized *T. leana* from CAS loc. 1294, Sand Flat, Redding area, Shasta Co., California, but these specimens are all *L. fitchi*. This locality is in Member I and probably Mid Turonian in age.

Fossil Localities

Descriptions for nearly half of the cited numbered localities have been previously published. Most of these are listed briefly with a reference. For abbreviations see page 4.

82 UCB: Mt. Diablo, Diablo quad., Contra Costa Co., California SW¼ sec. 8, T 1 S, R 1 E. Coll: ?, 1911 [Cenomanian?]
86 CIT: Santiago-Aliso Creek divide, Corona quad., Orange Co., California—Williams Fm., Pleasants ss. mem. [late Campanian] (Saul, 1974, p. 1093).
M175 USGS: Logan Ridge, Lodoga quad., Colusa Co., California. 950′N, 200′E of SW cor. sec. 33, T 18 N, R 4 W.—In calcareous fine- to medium-grained sandstone, "Antelope Shale" of Kirby, 75′-100′ below base of Venado Fm. [Turonian].
M176 USGS: Logan Ridge, Lodoga quad., Colusa Co., California. On W line sec. 33, 2100′S of NW cor. sec. 33, T 18 N, R 4 W.—In pebbly deformed mudstone, "Antelope Shale" of Kirby, ca. 40′ below base of Venado Fm. [Cenomanian].
M253 USGS: Boulder Creek, Wimer quad., Jackson Co., Oregon [late early to middle Albian] (Jones, 1960a, p. 155).
405 USGS: SE of Pentz, Cherokee quad., Butte Co., California—Chico Fm. [early Campanian] (Turner, 1894, p. 459).
445B CAS: Griffin Creek, Medford quad., Jackson Co., Oregon. Near Griffin Creek, west of Phoenix. Coll: F. M. Anderson—Hornbrook Fm. [Cenomanian or early Turonian].
590 CIT: right side Jalama Ranch Creek, 4.12 miles N 28½°E from B.M. at Concepcion, western Santa Ynez Range, Santa Barbara Co., California. Coll: J. Dorrance, Oct. 1929—Jalama Fm. [early Maestrichtian].

905   USGS: Cow Creek, Canyonville quad., Douglas Co., Oregon—Days Creek Fm. [Hauterivian] (Imlay, 1960, p. 187).

907   USGS: Grave Creek, Wimer quad., Jackson Co., Oregon [late early to middle Albian] (Jones, 1960a, p. 155).

1016   CIT: Chico Creek, Paradise quad., Butte Co., California—Chico Fm. [Santonian] (Matsumoto, 1960, p. 101).

1051   USGS: Texas Springs, Redding quad., Shasta Co., California. Sandstones SW of cabin at Texas Springs. Coll: Diller, Stanton, and Storrs, 1893—Budden Canyon Fm. [late early Albian].

1159   CIT: Dayton Canyon, Calabasas quad., Los Angeles Co., California [late Campanian] (Matsumoto, 1960, p. 103).

1169   CIT: Baker Creek, Corona quad., Orange Co., California. Crest of spur between forks of Baker Creek, just N of Pixley house, approx. 6000′ E of juncture of Baker and Santiago Creeks, Santa Ana Mts. Coll: W. P. Popenoe, May 16, 1935—Ladd Fm., top of Holz shale mem. *fide* Popenoe; Williams Fm., Schulz Ranch cgl. *fide* Schoellhamer [mid Campanian].

1232   CIT: Oak Run, Redding quad., Shasta Co., California. Right side of valley of E branch Oak Run, ¼ mi. N of S line sec. 2, 3350′ N 33°30′E from SW cor. sec.2, T 32 N, R 2 W. Coll: Popenoe and Ahlroth, July 4, 1936—In cross-bedded sandstones interbedded with conglomerate, Member V [Coniacian].

1245   USGS: near Riddle, Canyonville quad., Douglas Co., Oregon—Days Creek Fm. [Hauterivian] (Imlay, 1960, p. 177).

1277   HSU: West Coast Cedros Island, Baja California, Mexico. Arroyo Vargas, on bank about 400 m. from mouth. Coll: F. H. Kilmer—In hard, west-dipping sandstone, Valle Fm., Upper Member [Turonian?].

1294   CAS: Sand Flat, Redding quad., Shasta Co., California. 5 miles N of Redding—Member I [Turonian].

2166   USGS: Simmons Cut, 3 miles N of Waldo, Josephine Co., Oregon [Hauterivian or Barremian] (Imlay, 1960, p. 189).

2187   USGS: Jackson Creek, ½ mile SW of Jacksonville, Medford quad., Jackson Co., Oregon—Hornbrook Fm. [Cenomanian? or early Turonian?].

2190   USGS: see 907.

2365   CAS: Big Tar Canyon area, Garza Peak quad., Kings Co., California—Panoche Fm. [Campanian] (Saul, 1974, p. 1080).

2415   UCLA: Bee Canyon, El Toro quad., Orange Co., California—Williams Fm., Pleasants ss. mem. [late Campanian] (Matsumoto, 1960, p. 154).

2604   UCB: Elder Creek, Colyear Springs quad., Tehama Co., California. About 2¾ miles below Lowery in the gorge of Elder Creek, 500′ below 2nd falls below the suspension bridge, right ·bank of stream about 200′ below second cgl. (different fossiliferous blocks given different letters). Coll: E. L. Packard—Great Valley Series [Cenomanian? blocks in Turonian?].

3339   USGS: A deep cut (= Simmons Cut?), 4 mi. N of Waldo, Cave Junction quad., Josephine Co., Oregon [Hauterivian] (Imlay, 1960, p. 189).

3619   UCLA: Chico Creek, Paradise quad., Butte Co., California—Chico Fm. [Santonian] (Matsumoto, 1960, p. 155).

3621   UCLA: Chico Creek, Paradise quad., Butte Co., California. E of Chico Creek Co. road in upper part of meadow to N of old prune orchard, 2050′S, 2350′W of NE cor. sec.12, T 23 N, R 2 E, Coll: L. R. and R. B. Saul, Aug. 17, 1952—Chico Fm. [Santonian].

3623   UCLA: Chico Creek, Paradise quad., Butte Co., California—Chico Fm. [Santonian] (Matsumoto, 1960, p. 155).

3625   UCLA: Chico Creek, Paradise quad., Butte Co., California—Chico Fm. [Santonian] (Saul and Popenoe, 1962, p. 328).

3763   UCLA: North Fork Cottonwood Creek, Ono quad., Shasta Co., California—Bald Hills Fm. [Cenomanian] (Murphy and Rodda, 1960, p. 856).

3882   UCLA: S flowing tributary to Clear Creek N of Readings Bar, Redding quad., Shasta Co., California. About 100′ SE of SE cor. small pond at old placer mine, 1000′S, 450′W of NE cor. sec. 36, T 31 N, R 6 W. Coll: Peter Rodda, July, 1955—Budden Canyon Fm. [late early Albian].

3899   UCLA: S flowing tributary to Clear Creek N of Readings Bar, Anderson quad., Shasta Co., California. About 50 yds. S of UCLA 3882, W slope of low ridge and amongst piles of placer gravels, 1200′S, 450′W of NE cor. sec. 36, T 31 N, R 6 W. Coll: Peter Rodda, Aug. 1955—Budden Canyon Fm. [late early Albian].

3909   UCLA: Texas Springs area, Redding quad., Shasta Co., California. Quarry near Centerville-Olney Creek roads junction, 2100′N, 150′E of SW cor. sec. 28, T 31 N, R 5 W. Coll: Peter Rodda, Sept. 1955—Budden Canyon Fm. [late early Albian].

3915   UCLA: Texas Springs area Redding quad., Shasta Co., California. 1300′N, 1750′W of SE cor. sec. 29, T 31 N, R 5 W. Coll: Peter Rodda, Sept. 1955—Budden Canyon Fm. [late early Albian].

4082   UCLA: Tuscan Springs, Tuscan Springs quad., Tehama Co., California. On Little Salt Creek, about 10 miles NE of Red Bluff, near center NE¼ sec. 32, T 28 N, R 2 W. Coll: Popenoe, et al.—Chico Fm. [early Campanian].

4118   UCLA: Jalama Creek, Lompoc Hills quad., Santa Barbara Co., California—Jalama Fm., Member V [early Maestrichtian] (Dailey and Popenoe, 1966, fig. 1).

4122   UCLA: Jalama Creek, Lompoc Hills quad., Santa Barbara Co., California—Jalama Fm., Member V [early Maestrichtian] (Dailey and Popenoe, 1966, fig. 1).

4124   UCLA: Jalama Creek, Lompoc Hills quad., Santa Barbara Co., California—Jalama Fm., Member V [early Maestrichtian] (Dailey and Popenoe, 1966, fig. 1).

4127   UCLA: Jalama Creek, Lompoc Hills quad., Santa Barbara Co., California—Jalama Fm., Member V [early Maestrichtian] (Dailey and Popenoe, 1966, fig. 1).

4132   UCLA: Arroyo el Bulito, Lompoc Hills quad., Santa Barbara Co., California—Jalama Fm., Member V [early Maestrichtian] (Dailey and Popenoe, 1966, fig. 1).

4133   UCLA: Arroyo el Bulito, Lompoc Hills quad., Santa Barbara Co., California—Jalama Fm., Member V [early Maestrichtian] (Dailey and Popenoe, 1966, fig. 1).

4134   UCLA: Cañada de Santa Anita, Lompoc Hills quad., Santa Barbara Co., California—Jalama Fm., Member V [early Maestrichtian] (Dailey and Popenoe, 1966, fig. 1).

4138   UCLA: Cañada de Santa Anita, Lompoc Hills quad., Santa Barbara Co., California—Jalama Fm., Member V [early Maestrichtian] (Dailey and Popenoe, 1966, fig. 1).

4141   UCLA: Jalama Creek area, Lompoc Hills quad., Santa Barbara Co., California—Jalama Fm., Member III [early Maestrichtian] (Dailey and Popenoe, 1966, fig. 1).

4144   UCLA: Jalama Creek, Lompoc Hills quad., Santa Barbara Co., California—Jalama Fm., Member III [early Maestrichtian] (Dailey and Popenoe, 1966, fig. 1).

4250   UCLA: Young Ranch, Hornbrook quad., Siskiyou Co., California—Hornbrook Fm. [late Turonian] (Saul and Popenoe, 1962, p. 329).

4670   UCLA: Grave Creek, Wimer quad., Jackson Co., Oregon. W bank and bed of Grave Creek, W of road, 2400′N of SW cor. sec. 5, T 34 S, R 4 W. Coll: W. P. Popenoe, Aug. 20, 1960 [Albian].

5425   UCLA: Griffin Creek, Medford quad., Jackson Co., Oregon. Bank of irrigation canal on Lowell Ranch, E of Griffin Creek Rd. and approx. .7 mi. from intersection Pioneer and Griffin Creek Rds. Coll: T. Susuki, July 14, 1962—Hornbrook Fm. [Cenomanian? early Turonian?].

5427   UCLA: Griffin Creek area, Medford quad., Jackson Co., Oregon. Canal-cut at end of dirt road which is about .4 mi. S of intersection of Griffin Creek Rd. and South Stage Road. Coll: T. Susuki, July 14, 1962—Hornbrook Fm. [Cenomanian? early Turonian?].

M6003   USGS: Panoche de San Juan y los Carrisatillos Grant, Ortigalita Peak quad., Merced Co., California. 5,450′S, 42.55°E from Piedra Azul Spring, on SE flank of 1,625′ hill. Coll: John Dillon and Tom Dibblee, 1972—Great Valley Series [early Turonian].

6010   UCLA: Old Cow Creek, Whitmore quad., Shasta Co., California. Small tributary to Old Cow Creek, back of Schobbel house, 500′S, 800′W of NE cor. sec. 21, T 32 N, R 1 W. Coll: W. P. Popenoe, Oct. 31, 1969—Member V? [Late Coniacian].

6138   UCLA: Hagerdorn (Young) Ranch, Hornbrook quad., Siskiyou Co., California. 1400′NE of old Hagerdorn (Young) ranch house and 1700′N, 2000′W of SE cor. sec. 26, T 46 N, R 6 W. Coll: W. P. Popenoe, June 21, 1973—Hornbrook Fm. [late Turonian].

6140   UCLA: Rocky Gulch, Hornbrook quad., Siskiyou Co., California. N bank of Rocky Gulch, 1500′N, 700′W of SE cor. sec. 30, T 47 N, R 6 W, approx. 1.1 miles SW of center of Henley. Coll: W. P. Popenoe, Aug. 25, 1954—Hornbrook Fm., Member I, about 100-200′ above base [Turonian].

6168   UCLA: South of Antone, Dayville quad., Wheeler Co., Oregon. Along road E side of Rock Creek, At NW cor. SE¼ sec. 11, T 13 S, R 24 E. Coll: W. P. Popenoe and R. W. Imlay, 1959 [Cenomanian].

6177   UCLA: W side Cascade Bay, E side Harrison Lake, British Columbia—Brokenback Hill Fm. *fide* Crickmay, but possibly *Peninsula* Fm. [mid Valanginian] (Crickmay, 1930b, p. 42, loc. 47).

6275   UCLA: Lane Creek, Canyonville quad., Douglas Co., Oregon. Road cut N side Lane Creek, approx. 1750′S, 500′W of NE cor. sec. 15, at about n of Lane, T 30 S, R 6 W, 1½ miles N 50° W of Riddle. Coll: L. R., R. B., R. B., and R. L. Saul, Sept. 5, 1974—Days Creek Fm. [Hauterivian].

6276   UCLA: South Umpqua River at Days Creek, Days Creek quad., Douglas Co., Oregon. Massive sandstone bed N side E abutment bridge across South Umpqua River just N of mouth of Days Creek, SW cor. "sec. 47," 1500′N, 1000′W of NE cor. sec. 16, T 30 S, R 4 W. Coll: L. R., R. B., R. B., and R. L. Saul, Sept. 5, 1974—Days Creek Fm. [Hauterivian].

6298   UCLA: Los Gatos Creek, Coalinga quad., Fresno Co., California. Clast from road edge, 1st gully W of Nunez Canyon, Los Gatos Creek Rd., approx. 2400′S of NW cor. sec. 4, T 20 S, R 14 E. Coll: L. R. Saul, Dec. 13, 1971—Panoche Fm. [late Campanian].

6304   UCLA: Pine Timber Gulch, Whitmore quad., Shasta Co., California. Soft sandstone in channel of gulch, approx. 2000′N, 2400′E of SW cor. sec. 9, T 31 N, R 1 W. Coll: W. P. Popenoe and H. V. Church, Aug. 11, 1936—Member V? [Coniacian].

6305   UCLA: Klamath River, Hornbrook quad., Siskiyou Co., California. Near top of bluff on E side Klamath River, and just N of mouth of Osburner Gulch, approx. 1300′N, 500′E of SW cor. sec. 33, T 47 N, R 6 W. Coll: Monty Elliott and W. P. Popenoe, June, 1975—Hornbrook Fm. [Turonian].

6306   UCLA: Bellinger Hill, Medford quad., Jackson Co., Oregon. Bellinger Lane, nearly at crest of Bellinger Hill, about 1000′E of junction of Bellinger Lane with South Stage Road, and approx. 1¼ miles SE of Jacksonville. Coll: W. P. Popenoe, June 17, 1975—Hornbrook Fm. [Cenomanian].

6307   UCLA: Griffin Creek, Medford quad., Jackson Co., Oregon. Jess Baker farm, ½ mi. S of Griffin Creek schoolhouse, about ¼ mi. E of farmhouse and 3 miles SW of center of Medford. Coll: W. P. Popenoe, May 19, 1944—Hornbrook Fm. [Cenomanian or Turonian].

6310   UCLA: Topanga Canyon, Topanga quad., Los Angeles Co., California. Near top of cliff above Pacific Coast Highway, W side of gully approx. 200′S of Topanga Canyon, about 2400′S, 950′E of NW cor. sec. 32, T 1 S, R 16 W. Coll: John Alderson, Aug. 10, 1975 [late Campanian].

6312   UCLA: "Sagaser Hill", Garza Peak quad., Kings Co., California. NW trending ridge ½+ mile W of Sagaser Ranch House, Big Tar Canyon, about 40′ from top, 2000′N, 475′W of SE cor. sec. 13, T 23 S, R 16 E. Coll: L. R. Saul, Feb. 5, 1974—Panoche Fm. [Coniacian?].

6315   UCLA: Jacksonville, Medford quad., Jackson Co., Oregon. Sandstone just above basement on ridge along dirt road N 70° W of Jacksonville dump, approx. 1600′S, 2000′W of NE cor. sec. 5, T 38 S, R 2 W. Coll: L. R. and R. B. Saul, Aug. 24, 1975—Hornbrook Fm. [Cenomanian].

A-7159 UCB: Benicia, Benicia quad., Solano Co., California. Bulldozed road SE of burned house (650′S, 1275′W of NE cor.) sec. 34, T 3 N, R 3 W. Coll: Marion Ricks, 1966 [Turonian].

A-7601 UCB: N of Pescadero Point, Half Moon Bay quad., San Mateo Co., California. South of mouth of Pescadero Creek, soft gray sand bed in black silty shale along beach at small point nearly 1 mile N of Pescadero Point. Coll: J. W. Durham and students, Oct. 21, 1951—Pigeon Point Fm. [early Maestrichtian].

10117   USGS: South Fork of John Day River, Dayville quad., Grant Co., Oregon. About 6 miles S of Dayville, T 13 S, R 26 E. Coll: Campbell and Martin [Cenomanian].

22498   USGS: see 905 USGS.

23542   USGS: Boulder Creek, tributary to Grave Creek, Wimer quad., Jackson Co., Oregon [middle to early late Albian] (Jones, 1960a, p. 155).

26268   USGS: Five to six miles southwest of Dayville, Grant County, Oregon. Sec. 25 or 26, T 13 S, R 26 E. Coll: Zingula and Peeples, 1954 [Cenomanian].

27830   CAS: 4.9 miles E of Millville on road to Whitmore, Shasta Co., California. Center S side SW/4 sec. 33, T 32 N, R 2 W. Coll: Taff, Hanna, and Cross, May 2, 1934—Member V or VI [Santonian].

27838   CAS: Chico Creek, Paradise quad., Butte Co., California. Chico Fm. [early Campanian] (Matsumoto, 1960, p. 84).

28102   CAS: Basin Hollow, near Millville, eastern Shasta Co., California. Coll: Dana Russell, 1928 [Santonian].

29580   CAS: San Luis Ranch, Pacheco Pass quad., Merced Co., California. Concretion in pebbly sandstone, 1 mile WNW of San Luis Ranch—Panoche Fm. [Cenomanian].

31334   CAS: Along the shore, ½ mile N of Bolsa Point, Pigeon Point quad., San Mateo Co., California—Pigeon Point Fm. [early Campanian] (Saul and Popenoe, 1962, p. 329).

33704   CAS: Quinto Creek on Howard Ranch, Pacheco Pass quad., Merced Co., California—Panoche Fm. [late Campanian] (Matsumoto, 1960, p. 94).

33767   CAS: Days Creek, Days Creek quad., Douglas Co., Oregon. NE of Canyonville—Days Creek Fm. [Hauterivian].

LITERATURE CITED

ALLAN, JOYCE
  1959. Australian Shells. [2nd ed.] Melbourne, Georgian House, xxi + 487 pp., 44 pls., 112 text figs.
ALLISON, E. C.
  1955. Middle Cretaceous Gastropoda from Punta China, Baja California, Mexico. Jour. Paleontology, vol. 29, pp. 400-432, pl. 40-44, 3 text figs.
ALLISON, E. C., M. G. ACOSTA, D. L. FIFE, J. A. MINCH, and KATSUO NISHIKAWA [eds.]
  1970. Pacific slope geology of northern Baja California and adjacent Alta California. Amer. Assoc. Petrol. Geol., Soc. Econ. Paleon. and Min., and Soc. Explor. Geophys., Pacific Sections, Guide-book Fall Field Trip, 160 pp.
ANDERSON, F. M.
  1902. Cretaceous deposits of the Pacific Coast. Calif. Acad. Sci., Proc., 3rd ser., vol. 2, pp. 1-154, pls. 1-12.
  1958. Upper Cretaceous of the Pacific Coast. Geol. Soc. Amer., Mem. 71, 378 pp., 75 pls.
BAKER, C. L.
  1927. Exploratory geology of a part of southwestern Trans-Pecos Texas. Univ. Texas Bull. 2745, 70 pp., 1 pl.
BOLTON, T. E.
  1965. Catalogue of type invertebrate fossils of the Geological Survey of Canada. Volume II. Canada Geol. Surv., 344 pp.
BRETSKY, P. W.
  1973. Evolutionary patterns in the Paleozoic Bivalvia: documentation and some theoretical considerations. Geol. Soc. America. Bull. v. 84, pp. 2079-2096, 5 text figs.
BROWN, R. D., JR., and E. I. RICH
  1960. Early Cretaceous fossils in submarine slump deposits of Late Cretaceous age, northern Sacramento Valley, California. U.S. Geol. Surv., Prof. Paper 400B, pp. B318-B320.
CALIFORNIA STATE MINING BUREAU
  1899. Catalogue of the State Museum of California. Vol. 5 being the collection made by the State Mining Bureau from September 1, 1890 to March 30, 1897. Sacramento, 292 pp.
CASEY, RAYMOND
  1952. Some genera and subgenera, mainly new, of Mesozoic heterodont lamellibranchs. Malacol. Soc. London, Proc., vol. 29, pp. 121-176, pls. 7-9, 100 text figs.
CHAMBERLAIN, J. A.
  1976. Flow patterns and drag coefficients of cephalopod shells. Palaeontology, v. 19, pp. 539-563, 16 text figs.
CHAMBERLAIN, J. A., and G. E. G. WESTERMANN
  1976. Hydrodynamic properties of cephalopod shell ornament. Paleobiology, v. 2, pp. 316-331, 13 figs.
COATES, J. A.
  1974. Geology of the Manning Park Area, British Columbia. Canada Geol. Surv., Bull. 238, 177 pp., 9 tabs., 12 figs., 21 pls., A6 tabs.
COX, L. R.
  1952. Notes on the Trigoniidae, with outlines of a classification of the family. Malacol. Soc. London, Proc. vol. 29, pp. 45-70, pls. 3-4.
CRAGIN. F. W.
  1893. A contribution to the invertebrate paleontology of the Texas Cretaceous. Texas Geol. Surv., 4th Ann. Rept., Pt. 2, pp. 139-246, pls. 24-46.
  1905. Paleontology of the Malone Jurassic Formation of Texas. U.S. Geol. Surv., Bull. 266, 172 pp., 29 pls.
CRICKMAY, C. H.
  1930a. The Jurassic rocks of Ashcroft, British Columbia. Univ. Calif. Publ., Dept. Geol. Sci., Bull., vol. 19, pp. 23-74, pls. 2-7, 1 map.
  1930b. Fossils from Harrison Lake Area, British Columbia. Canada Nat. Mus., Bull. 63, Geol. Ser. 51, pp. 33-66, 82-113, pls. 8-23, 7 text figs.

1932. Contributions toward a monograph of the Trigoniidae, I. Amer. Jour. Sci., vol. 224, pp. 443-464, 2 pls.

1962. Gross stratigraphy of Harrison Lake area, British Columbia. Calgary, Evelyn deMille Books (publ. by author), Art. 8, pp. 1-12, 1 pl.

DAILEY, D. H., and W. P. POPENOE

1966. Mollusca from the Upper Cretaceous Jalama Formation, Santa Barbara County, California. Univ. Calif. Publ. Geol. Sci., vol. 65, 41 pp., 6 pls., 3 text figs.

DEECKE, W.

1925. Trigoniidae mesozoicae (Myophoria exclusa). *In* Diener, C. [ed.] Fossilium catalogus, I: Animalia, W. Junk, Berlin, 306 pp.

1926. Über die Trigonien. Palaeontologische Zeitschrift, vol. 7, pp. 65-101.

DIETRICH, W. O.

1933a. Zur Stratigraphie und Palaeontologie der Tendaguruschichten. Palaeontographica, Sup. 7 [Wissenschaftliche Ergebnisse der Tendaguru-Expedition 1909-1912], Ser. 2, Teil 2, pp. 1-86, pls. 1-12, 1 text fig.

1933b. Das Muster der Gattung *Trigonia* (Moll. Lam.). Gesells. Naturf. Freunde Berlin, Sitzungsberichte, pp. 326-332.

FREY, R. W. and J. D. HOWARD

1970. Comparison of Upper Cretaceous ichnofaunas from siliceous sandstones and chalk, Western Interior Region, U.S.A., *In* T. P. Crimes and J. C. Harper [eds.]. Trace fossils. Geol. Jour. spec. issue 3, pp. 141-166, text figs. 1-8, tabs. 1-3.

FLEMING, C. A.

1964. History of the bivalve family Trigoniidae in the South-west Pacific. Australian Jour. Sci., vol. 26, pp. 196-204, 13 text figs.

GABB, W. M.

1864. Description of the Cretaceous fossils. Calif. Geol. Surv., Paleontology, vol. 1, pp. 57-243, 1844, pls. 9-32, 1865.

1877. Notes on American Cretaceous fossils, with descriptions of some new species. Acad. Nat. Sci. Phila., Proc., vol. 28, pp. 276-324, pl. 17.

GILLET, SUZETTE

1924-25. Études sur les lamellibranches néocomiens. Soc. Géol. France, (5) Mém. 3, pp. 1-224, 2 pls., 95 text figs., 1924; pp. 225-339, 4 maps. 1925.

1965. Les trigonies du Crétacé inférieur. France Bur. Rech. Géol. Min., Mém. 34 [Colloque sur le Crétacé inférieur, Lyon, septembre, 1963], pp. 399-407, 2 text figs.

GORDON, W. A.

1973. Marine life and ocean surface currents in the Cretaceous. Jour. Geology, vol. 81, pp. 269-284, 6 text figs.

GOULD, S. J.

1969. The byssus of trigonian clams: phylogenetic vestige or functional organ? Jour. Paleontology, vol. 43, pp. 1125-1129, 2 text figs.

GOULD, S. J. and C. C. JONES

1974. The pallial ridge of *Neotrigonia:* functional siphons without mantle fusion. Veliger, vol. 17, pp. 1-7, 1 pl., 2 text figs.

HALL, C. A., JR.

1975. Latitudinal variation in shell growth patterns of bivalve molluscs: implications and problems. *In* G. D. Rosenberg and S. K. Runcorn [eds.], Growth rhythms and the history of the earth's rotation. New York and London, John Wiley & Sons, pp. 163-175, 6 text figs.

HALL, C. A., JR., D. L. JONES, and S. A. BROOKS

1959. Pigeon Point Formation of Late Cretaceous age, San Mateo County, California. Amer. Assoc. Petroleum Geologists, Bull. vol. 43, pp. 2855-2859, 2 text figs.

IMLAY, R. W.

1952. Correlation of the Jurassic Formations of North America, exclusive of Canada. Geol. Soc. Amer. Bull., vol. 63, pp. 953-992, 2 pls., 4 text figs.

1960. Ammonites of Early Cretaceous age (Valanginian and Hauterivian) from the Pacific Coast States. U. S. Geol. Surv., Prof. Paper 334-F, pp. iii + 167-228, pls. 24-43, figs. 34-36, tabs. 1-4, chart 1.

　　1964. Marine Jurassic pelecypods from central and southern Utah. U. S. Geol. Surv., Prof. Paper 483-C, 42 pp., 4 pls.

JELETZKY, J. A.

　　1965. Late Upper Jurassic and early Lower Cretaceous fossil zones of the Canadian western Cordillera, British Columbia. Canada Geol. Surv., Bull. 103, 70 pp., 22 pls.

JONES, D. L.

　　1960a. Lower Cretaceous (Albian) fossils from southwestern Oregon and their paleogeographic significance. Jour. Paleontology, vol. 34, pp. 152-160, pl. 29, 2 text figs.

　　1960b. Pelecypods of the genus *Pterotrigonia* from the west coast of North America. Jour. Paleontology, vol. 34, pp. 433-439, pls. 59-60, 2 text figs.

KAUFFMAN, E. G.

　　1975. Dispersal and biostratigraphic potential of Cretaceous benthonic Bivalvia in the Western Interior. Geol. Assn. Canada, Sp. Paper 13, pp. 163-194, 4 text figs.

KAUFFMAN, E. G., and N. F. SOHL

　　1974. Structure and evolution of Antillean Cretaceous rudist frameworks. Verhandl. Naturf. Ges. Basel, vol. 84, pp. 399-467, 27 text figs.

KITCHIN, F. L.

　　1903. The Jurassic fauna of Cutch. The Lamellibranchiata. Genus *Trigonia*. Palaeont. Indica, ser. 9, vol. 3, pt. 2, no. 1, 122 pp., 10 pls.

　　1913. The invertebrate fauna and palaeontological relations of the Uitenhage Series. So. African Mus. Annals, vol. 7, pp. 21-250, pls. 2-11.

KOBAYASHI, TEIICHI, and MASAHISA AMANO

　　1955. On the Pseudoquadratae trigonians, *Steinmanella*, in the Indo-Pacific Province. Japan. Jour. Geology Geography, vol. 26, pp. 195-208, pls. 13-15.

KOBAYASHI, TEIICHI, KAZUO MORI, and MINORU TAMURA

　　1959. The bearing of the trigoniids on the Jurassic stratigraphy of Japan. Studies on the Jurassic trigonians in Japan, VIII. Japan. Jour. Geology Geography, vol. 30, pp. 273-292, 5 tables.

KOBAYASHI, TEIICHI, and MINORU TAMURA

　　1955. The Myophorellinae from North Japan. Studies on the Jurassic trigonians in Japan, IV. Japan. Jour. Geology Geography, vol. 26, p. 89-104, pl. 5-6.

LUYENDYK, B. P., DONALD FORSYTH, and J. S. PHILLIPS

　　1972. Experimental approach to the paleocirculation of the oceanic surface waters. Geol. Soc. Amer., Bull. vol. 83, pp. 2649-2664, 6 figs.

LYCETT, JOHN

　　1872-83. A monograph of the British fossil Trigoniae. Palaeontograph. Soc. London, 245 + 19 pages, 41 + 4 pls.

MACKENZIE, J. D.

　　1916. Geology of Graham Island, British Columbia. Canada Geol. Surv., Mem. 88, viii + 221 pp., 16 pls., maps 176A, 177A.

MATSUMOTO, TATSURO

　　1960. Upper Cretaceous ammonites of California, Part III. Kyushu Univ., Mem. Fac. Sci., Ser. D, Geology, Sp. vol. 2, 204 pp., 20 text figs., 2 pls.

MC ALESTER, A. L.

　　1965. Life habits of the "Living Fossil" bivalve *Neotrigonia*. Geol. Soc. Amer. Program, Ann. Meet. p. 102.

MC LEARN, F. H.

　　1949. Jurassic formations of Maude Island, and Alliford Bay, Skidegate Inlet, Queen Charlotte Islands, British Columbia. Canada Geol. Surv., Bull. 12, v + 19 pp., 3 figs.

　　1972. Ammonoids of the Lower Cretaceous sandstone member of the Haida Formation, Skidegate Inlet, Queen Charlotte Islands, western British Columbia. Canada Geol. Surv., Bull. 188, 78 pp., 45 pls., 3 figs.

MERRIAM, C. W.

　　1941. Fossil turritellas from the Pacific Coast region of North America. Univ. Calif. Publ., Bull. Dept. Geol. Sci., vol. 26, pp. 1-214, pls. 1-41, 19 text figs., 1 map.

MERRIAM, J. C.
    1895. A list of type specimens in the Geological Museum of the University of California, which have
        served as originals for figures and descriptions in the Paleontology of the State Geological Survey
        of California under J. D. Whitney. Compiled for the use of workers in California geology. Univ.
        Calif., Berkeley, Geol. Dept. (3 printed unnumbered pages). Reprinted by A. W. Vogdes, 1896,
        Calif. State Mining Bur., Bull. 10, pp. 21-23; 2nd ed. 1904, Bull. 30, pp. 39-42.
MOORE, R. C. [ed.]
    1969. Treatise on invertebrate paleontology. Lawrence, Kansas: Kansas Univ. Press and Geol. Soc.
        Amer., Pt. N. Mollusca 6, Bivalvia 1-2: 952 pp., figs. 1-103, A1-10, B1-5, C1-107, D1-76, E1-276,
        F1-32, G1, H1-2.
MULLER, J. E., and J. A. JELETZKY
    1970. Geology of the Upper Cretaceous Nanaimo Group, Vancouver Island and Gulf Islands, British
        Columbia. Canada Geol. Surv., Paper 69-25, vi + 77 pp., 3 tables, 11 figs.
MURPHY, M. A.
    1969. Geology of the Ono Quadrangle, Shasta and Tehama Counties, California. Calif. Div. Mines and
        Geology, Bull. 192, 28 pp., 1 pl., 3 figs., 10 photos.
MURPHY, M. A., and P. U. RODDA
    1960. Mollusca of the Cretaceous Bald Hills Formation of California. Jour. Paleontology, vol. 34,
        pp. 835-858, pls. 101-107, 2 text figs.
NAKANO, MITSUO
    1960. Stratigraphic occurrences of the Cretaceous trigoniids in the Japanese Islands and their faunal
        significances. Hiroshima Univ., Jour. Sci., (C), vol. 3, pp. 215-279, pls. 23-30, 19 tables.
    1963. On the Rutitrigoniinae. Hiroshima Univ., Geol. Rept. 12, pp. 513-529, pl. 56, 1 table.
    1968. On the Quadratotrigoniinae. Japan. Jour. Geol. and Geog., vol. 39, pp. 27-41.
    1970. Considerations on the life mode of *Nipponitrigonia* and *Pterotrigonia* from the Lower Cretaceous
        in Katsuuragawa Basin, Shikoku, Japan. Hiroshima Inst. Tech., Res. Bull. vol. 5, pp. 15-21,
        2 text figs. (in Japanese).
    1974. A new genus *Mediterraneotrigonia* nov. Hiroshima Inst. Tech. Res. Bull. vol. 9, pp. 77-80.
NEWELL, N. D., and D. W. BOYD
    1975. Parallel evolution in early trigoniacean bivalves. Amer. Mus. Nat. Hist. Bull., vol. 154, art. 2,
        pp. 55-162, 98 figs.
PACKARD, E. L.
    1916. Faunal studies in the Cretaceous of the Santa Ana Mountains of Southern California. Univ.
        Calif. Publ., Bull. Dept. Geology, vol. 9, pp. 137-159.
    1921. The Trigoniae from the Pacific Coast of North America. Univ. Oregon Publ., vol. 1, no. 9, 35 pp.,
        11 pls.
PECK, D. L., R. W. IMLAY, and W. P. POPENOE
    1956. Upper Cretaceous rocks of parts of southwestern Oregon and northern California. Amer. Assoc.
        Petrol. Geol., Bull., vol. 40, pp. 1968-1984.
PERRILLIAT-MONTOYA, M. C.
    1968. Fauna del Cretacico y del Terciario del Norte de Baja California. Paleontologia Mexicana, no. 25,
        36 pp., 8 pls.
POPENOE, W. P.
    1942. Upper Cretaceous formations and faunas of Southern California. Amer. Assoc. Petrol. Geol.,
        Bull. vol. 26, pp. 162-187.
POPENOE, W. P., R. W. IMLAY, and M. A. MURPHY
    1960. Correlation of the Cretaceous Formations of the Pacific Coast (United States and Northwestern
        Mexico). Geol. Soc. America, Bull., vol. 71, pp. 1491-1540, 1 pl.
RODDA, P. U.
    1959. Geology and paleontology of a portion of Shasta County, California. Univ. Calif., Los Angeles,
        unpubl. Ph.D. dissertation, 205 pp., 20 pls.
SANBORN, A. F.
    1960. Geology and paleontology of the southwest quarter of the Big Bend Quadrangle, Shasta County,
        California. Calif. Div. Mines, Spec. Rept. 63, 26 pp., 2 pls.

SAUL, L. R.

   1973. Evidence for the origin of the Mactridae (Bivalvia) in the Cretaceous. Univ. Calif. Publ. Geol.
          Sci., v. 97, 59 pp., 3 pls., 8 text figs.
   1974. Described or figured West Coast species of *Cymbophora*. Jour. Paleontology, vol. 48, pp. 1068-
          1095, 3 pls., 7 text figs.

SAUL, L. R., and W. P. POPENOE

   1962. *Meekia,* enigmatic Cretaceous pelecypod genus. Univ. Calif. Publ. Geol. Sci., vol. 40, pp. 289-344,
          6 pls., 4 text figs.

SAVELIEV, A. A.

   1958. Nizhnemelovye trigoniidy Mangyshlaka i Zapadnoi Turkmenii (s ocherkom sistematiki i filogenii
          semeistva). Leningrad, Gos. nauchnotekhn, izd-vo neftianoi i gorno-topliunoi litry, Leningrad-
          skoe ofd-nie (Trudy Vsesoueznogo neftianogo nauchno-issle-dovatel'skogo Geol. Inst.
          [VNIGRI]), vol. 125, pp. 1-516, 58 pls.

SEILACHER, ADOLF

   1972. Divaricate patterns in pelecypod shells. Lethaia, vol. 5, pp. 325-343, 18 text figs.

STALDER, WALTER

   1940. History of exploration and development of gas and oil in northern California. Calif. Div. Mines,
          Bull. 118, pp. 75-80.

STANLEY, S. M.

   1968. Post-Paleozoic adaptive radiation of infaunal bivalve mollusca—a consequence of mantel fusion
          and siphon formation. Jour. Paleontology, vol. 42, pp. 214-229, 13 text figs.
   1970. Relation of shell form to life habits in the Bivalvia (Mollusca). Geol. Soc. America, Mem. 125,
          296 pp., 40 pls., 47 text figs.

STEHLI, F. G., A. L. MC ALESTER, and C. E. HELSLEY

   1967. Taxonomic diversity of Recent bivalves and some implications for geology. Geol. Soc. America,
          Bull. vol. 78, pp. 455-466, 10 text figs.

STEWART, R. B.

   1930. Gabb's California Cretaceous and Tertiary type lamellibranchs. Acad. Nat. Sci. Phila., Spec.
          Publ. 3, 314 pp., 17 pls.

STOYANOW, ALEXANDER

   1949. Lower Cretaceous stratigraphy in Southeastern Arizona. Geol. Soc. America. Mem. 38, 169 pp.,
          27 pls., 2 text figs.

SUTHERLAND BROWN, A.

   1968. Geology of the Queen Charlotte Islands, British Columbia. Brit. Columbia Dept. Mines Petrol.
          Res., Bull. 54, 226 pp., 21 tabs., 45 figs., 18 pls.

TEVESZ, M. J. S.

   1975. Structure and habits of the 'living fossil' pelecypod *Neotrigonia*. Lethaia, vol. 8, pp. 321-327,
          4 text figs.

TURNER, H. W.

   1894. The rocks of the Sierra Nevada. U.S. Geol. Survey, 14th Ann. Rept., pt. 2, pp. 435-495, pls. 48-59.

VALENTINE, J. W.

   1971. Plate tectonics and shallow marine diversity and endemism, an actualistic model. System. Zool-
          ogy, v. 20, pp. 253-264, 4 text figs., 1 tab.

VALENTINE, J. W., and E. M. MOORES

   1972. Global tectonics and the fossil record. Jour. Geology, v. 80, pp. 167-186, 6 text figs.

WEAVER, C. E.

   1931. Paleontology of the Jurassic and Cretaceous of West Central Argentina. Univ. Washington,
          Mem. vol. 1, 594 pp., 62 pls.

WHITEAVES, J. F.

   1876. On some invertebrates from the coal-bearing rocks of the Queen Charlotte Islands. Canada
          Geol. Surv., Mesozoic Fossils, vol. 1, pt. 1, pp. 1-92, pls. 1-10.
   1879. On the fossils of the Cretaceous rocks of Vancouver and adjacent islands in the Strait of Georgia.
          Canada Geol. Surv., Mesozoic Fossils, vol. 1, pt. 2, pp. 93-190, pls. 11-20.

WHITNEY, MARION
   1952. Some Pelecypoda from the Glen Rose Formation of Texas. Jour. Paleontology, vol. 26, pp. 697-707, pls. 86-89.
WILBUR, K. M., and GARETH OWEN
   1964. Growth. *In* K. M. Wilbur and C. M. Yonge [eds.], Physiology of Mollusca. New York and London, Academic Press, Chapter 7, pp. 211-242, 14 text figs.

PLATES

PLATE 1

Yaadia jonesi n. sp.

Fig. 1. Latex pull from rock mold of syntype, left valve, × ¾, UCLA cat. no. 38568 from UCLA loc. 6276, showing beak, corcelet, and noded ribbing of escutcheon.

Fig. 2. Latex pull from rock mold of hypotype, left valve, × 1, USNM cat. no. 241669 from USGS loc. 3339, a juvenile specimen that has not yet developed the space between the anterior row of nodes and the oblique flank ribs.

Fig. 3. Latex pull from rock mold of syntype, right valve, × ¾, UCLA cat. no. 38567 from UCLA loc. 6276 on same block of sandstone and adjacent to UCLA cat. no. 38568, showing oblique flank ribs distinctly separated from row of large nodes along the anterior angulation.

Fig. 4-6. Paratype, left valve, UCLA cat. no. 38566 from UCLA loc. 6275; fig. 4, rock mold of exterior, × 1; fig. 5, latex pull of exterior, × 2, showing juvenile ribbing; fig. 6, rock mold of interior, × 2, showing impressions of "lateral teeth" at dorso-posterior corner, impressions of internal ribs delimiting in and out current flow areas, and the salient projection between the "cardinal" teeth that indicates the attachment of the pedal elevator muscle (see also Fleming, 1964, p. 198, fig. 5).

Fig. 7. Paratype, left valve, × ¾, USNM cat. no. 241670 from USGS loc. 22498, nearly complete outline of adult valve.

Fig. 8. Paratype, left valve, × ¾, USNM cat. no. 241671 from USGS loc. 1245, escutcheon and corcelet.

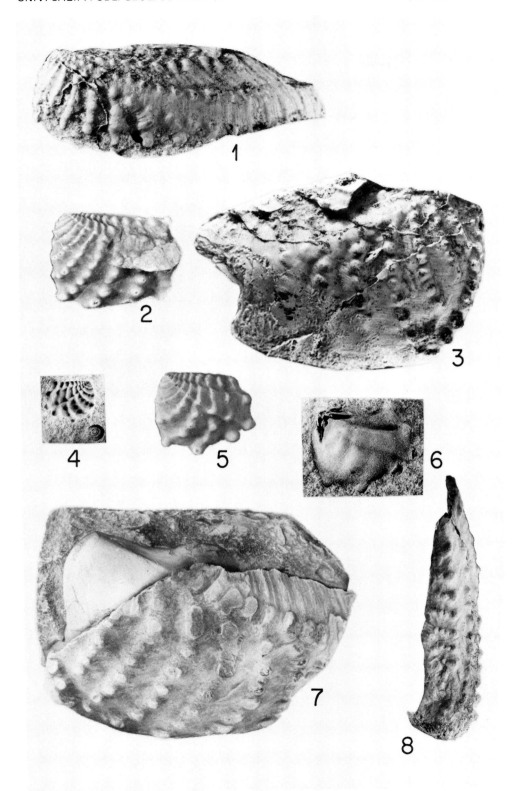

PLATE 2

All figures × ¾

Fig. 1-2. *Yaadia jonesi* n. sp., paratypes, left valves from USGS loc. 1245; fig. 1, USNM cat. no. 241671; fig. 2, USNM cat. no. 241672.

Figs. 3-5. *Yaadia whiteavesi* (Packard); fig. 3, plaster cast of holotype (CGS cat. no. 4997 from Queen Charlotte Islands, British Columbia), left valve, UCLA cat. no. 32195, separation of anterior row of nodes has been obscured by crushing of the specimen; fig. 4, hypotype, right valve, USNM cat. no. 241673 from USGS loc. 1051; fig. 5, plaster cast of paratype (CGS cat. no. 4997a from Queen Charlotte Islands, B.C.) left valve, UCLA cat. no. 32196, Anderson's (1958, p. 111) candidate for holotype.

PLATE 3

Yaadia whiteavesi (Packard)

Fig. 1. Latex pull from rock mold of hypotype, left valve, × 1, UCLA cat. no. 38575 from UCLA loc. 4670.

Fig. 2. Latex pull from rock mold of hypotype, right valve, × ¾, USNM 241674 from USGS loc. M 253.

Figs. 3-4. Holotype of *Trigonia perrinsmithi* Anderson, left valve, LSJU cat. no. 8697 from near Horsetown, Shasta Co., California, photos by Takeo Susuki, the anterior end is not crushed or broken and Anderson's (1958, pl. 2, fig. 7) reconstruction is incorrect; fig. 3, × .84, flank; fig. 4, × ½, beak view showing truncated anterior profile and corcelet.

Figs. 5-6. Left valves from UCLA loc. 4670; fig. 5, latex pull from rock mold of hypotype, × ¾, UCLA cat. no. 38576; fig. 6, rock mold of interior of hypotype, × 1, UCLA, cat. no. 38836, showing adductor muscle scars, pallial line and the salient projection between the hinge teeth that indicates the attachment of the pedal elevator muscle.

PLATE 4

Yaadia leana (Gabb)

Fig. 1. Holotype, left valve × 2, UCBMP cat. no. 12171 from Martinez, Contra Costa Co., California, photo by Takeo Susuki.

Fig. 2, 4. Hypotype, right valve, USNM cat. no. 241676 from USGS loc. 10117; fig. 2, × 1; fig. 4, × 2, beak view showing juvenile sculpture.

Fig. 3. Latex pull from rock mold of hypotype, right vavle × 2, CAS cat. no. 57980 from CAS loc. 29580.

5. Holotype of *Trigonia colusaensis* Anderson, right valve, × 1, CAS cat. no. 10632 from near Sites, Colusa Co., California, photo by Takeo Susuki.

Fig. 6. Hypotype, left valve, × 1, USNM 241677 from USGS loc. 26268, the anterior row of nodes is not prominent but is recognizable on a specimen of this size.

Figs. 7, 9. Hypotype, × 1, UCLA cat. no. 38716 from UCLA loc. 6306; fig. 7, right valve, compare with Pl. 5, fig. 4 which has more irregular sculpture on the anterior; fig. 9, beak view with more mature sculpture than in pl. 4, fig. 4.

Fig. 8. Holotype of *Trigonia wheelerensis* Anderson, left valve, × 1, UO cat. no. 26909 from Rock Creek, Wheeler Co., Oregon, photo by Takeo Susuki.

PLATE 5

Figs. 1-4. *Yaadia leana* (Gabb); fig. 1, paralectotype?, left valve, × ¾, UCBMP cat. no. 14498 from UCB loc. A-6458; fig. 2, paralectotype?, rock mold of posterior portion of valve including part of flank and corcelet, × ¾ UCBMP cat. no. 14499 from UCB loc. A-6458. This and fig. 1 may have been the basis for Gabb's (1864, pl. 31) fig. 262; and the four curious pustules on the growth lines of Gabb's figure could have been suggested by the nodes along the excutcheon of UCBMP 14499 (fig. 2); fig. 3, hypotype, right valve, × 1, USNM cat. no. 241679 from USGS loc. 2187, a speicmen upon which the anterior row of nodes can not be distinguished; fig. 4, hypotype, right valve, × 1, UCLA cat. no. 38715 from UCLA loc. 6306.

Figs. 5-8. *Yaadia* cf. *Y. californiana* (Packard); figs. 5-6, hypotype left valve, × 1, UCLA cat. no. 38602 from UCLA loc. 6138; fig. 5, flank; fig. 6, beak view showing juvenile ribbing on corcelet; fig. 7, plaster cast made by Joe Peck of a specimen from near Benicia, Solano Co., California, retained by its collector M. Ricks, × ¾, UCLA cat. no. 48649 [= UCBMP cat. no. 14500] from UCB loc. A-7159; fig. 8, hypotype, left valve, × 1, UCLA cat. no. 38603 from UCLA loc. 6138.

PLATE 6

All specimens × ¾

Figs. 1-2. *Yaadia* cf. *Y. californiana* (Packard), hypotype, UCLA cat. no. 38601 from UCLA loc. 6138; fig. 1, right valve; fig. 2, left valve.

Figs. 3-6. *Yaadia pinea* n. sp., from UCLA loc. 6304; fig. 3, holotype, right valve, UCLA cat. no. 38604; figs. 4-6, paratypes, left valves; fig. 4, UCLA cat. no. 38609; fig. 5, UCLA cat. no. 38605; fig. 6, UCLA cat. no. 38606, the specimen is somewhat compressed in the height direction.

PLATE 7

Figs. 1-2. *Yaadia pinea* n. sp., paratype, left valve, × 1, UCLA cat. no. 38608 from UCLA loc. 6304; fig. 1, beak view showing some corcelet sculpture; fig. 2, flank.

Figs. 3-10. *Yaadia branti* n. sp.; fig. 3, paratype, left valve, × 1, UCLA cat. no. 38614 from UCLA 3619, interior of juvenile showing "lateral teeth" at dorso-posterior corner and internal rib opposite the external median groove; fig. 4, paratype, left valve, × 1, UCLA cat. no. 38619 from CIT loc. 1016, specimen partially peeled but shows juvenile sculpture; figs. 5, 8, holotype, right valve, × ¾, UCLA cat. no. 38613 from UCLA 3619; fig. 5, flank, nodes have coalesced less on this specimen than on those figured in 7 and 9; fig. 8, beak view showing anterior profile and ornament of corcelet; figs. 6-7, paratype, left valve, UCLA cat. no. 38625 from UCLA 3625; fig. 6, × 1, anterior end; fig. 7, × ¾, flank, nodes coalesced along ribs; fig. 9, paratype, left valve, × ¾, UCLA 38623 from UCLA loc. 3623, nodes are strongly coalesced in growth line direction; fig. 10, paratype, left valve, × 1, UCLA cat. no. 38620 from CIT loc. 1016.

PLATE 8

Fig. 1. *Yaadia branti* n. sp., paratype, left valve, × ¾, UCLA cat. no. 38618 from CIT loc. 1016.

Figs. 2-7. *Yaadia tryoniana* (Gabb); fig. 2, hypotype and topotype, left valve, × 2, UCLA loc. 4082; fig. 3, hypotype, right valve, × 2, CAS cat. no. 57981 from CAS loc. 2365; figs. 4, 6, hypotype, left valve, × ¾, USNM cat. no. 241680 from USGS loc. 405; fig. 4, flank; fig. 6, beak view showing corcelet sculpture; fig. 5, hypotype, × ¾, UCLA cat. no. 38632 from CIT loc. 1169; fig. 7, hypotype, left valve, × ¾, UCLA cat. no. 38630 from CIT loc. 1169.

PLATE 9

Fig. 1. *Yaadia tryoniana* (Gabb), hypotype, right valve, × ¾, UCLA cat. no. 38631 from CIT loc. 1169.

Figs. 2-4, 6. *Yaadia robusta* n. sp., left valves; figs. 2-3, holotype, UCLA cat. no. 38634 from CIT loc. 1159; fig. 2, × 1, beak view showing riblets crossing the escutcheon; fig. 3, × ¾, flank, fig. 4, paratype, × ¾, UCLA cat. no. 38756 from UCLA loc. 6310; fig. 6, hypotype, × ¾, UCLA cat. no. 38636 from UCLA loc. 6298.

Fig. 5. *Yaadia* cf. *Y. robusta* Saul, hypotype, × ¾, UCLA cat. no. 38637 from UCLA 2415, beak view of double valved specimen showing anterior profile and weak sculpture on the corcelet.

PLATE 10

Fig. 1. *Yaadia* cf. *Y. robusta* Saul, hypotype, left valve, × ¾, UCLA cat. no. 38637 from UCLA loc. 2415.

Figs. 2-9. *Yaadia hemphilli* (Anderson); figs. 2, 5, hypotype, left valve, UCLA cat. no. 38651 from UCLA loc. 4124; fig. 2, × 1; fig. 5, × 2, beak view; figs. 3-4, hypotype, × 1, UCLA cat. no. 44424 from UCLA loc. 4118; fig. 3, right valve; fig. 4, left valve, note sculpture variation between two valves of same individual; fig. 6, hypotype, left valve, × 1, UCLA cat. no. 38640 from UCLA loc. 4144, beak view; fig. 7, hypotype, right valve, × 1, UCLA cat. no. 38662, from UCLA loc. 4134; figs. 8-9, holotype, × 1, CAS cat. no. 994 from "near Pescadero, California," photos by Takeo Susuki; fig. 8, beak view with ligament in place; fig. 9, right valve, flank.

PLATE 11

Fig. 1. *Yaadia jonesi* n. sp., latex pull from rock mold of hypotype, left valve, × 1, USNM cat. no. 241669 from USGS loc. 3339.

Fig. 2. *Yaadia leana* (Gabb), hypotype, left valve, × 1, USNM cat. no. 241677 from USGS loc. 26268.

Fig. 3. *Yaadia hemphilli* (Anderson), hypotype, right valve, × ¾, UCLA cat. no. 38663 from UCLA loc. 4127, compare elongate shape of mature valve to more equant shape of immature series figured on plate 10, figs. 2-4, 7.

Fig. 4. *Yaadia* cf. *Y. californiana* (Packard), hypotype, left valve, × 2 UCLA cat. no. 38603 from UCLA loc. 6138.

Fig. 5. *Yaadia branti* n. sp., paratype, left valve, × 2, UCLA cat. no. 38619, from CIT loc. 1016.

Fig. 6. *Yaadia hemphilli* (Anderson) hypotype, both valves, × 1, UCLA cat. no. 44424 from UCLA loc. 4118. Compare beak and corcelet sculpture of figs. 1-2, 4-6 to that of *Steinmanella transitoria quintucoensis* (Weaver) in fig. 8.

Figs. 7-8. *Steinmanella transitoria quintucoensis* (Weaver), hypotype, × ¾, UCR cat. no. 4653/1 from Weaver's loc. 1228 west central Argentina of late Mid Valanginian age; fig. 7, right valve; fig. 8, beak view showing anterior profile, juvenile sculpture and sculpture of corcelet and escutcheon.

Figs. 9, 11. *Quadratotrigonia* sp. B, hypotype, right valve × ¾, UCLA cat. no. 38837 from HSU loc. 1277; fig. 9, beak and anterior end; fig. 11, flank.

Fig. 10. *Quadratotrigonia* sp. A, hypotype, left valve, × ¾, UCLA cat. no. 38764 from UCLA loc. 3882, flank.

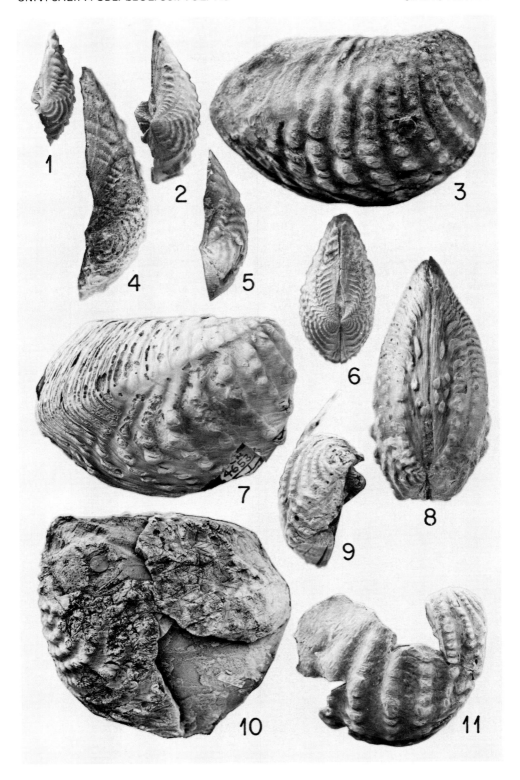

PLATE 12

Litschkovitrigonia? fitchi (Packard)

Fig. 1. Holotype of *Trigonia branneri* Anderson, left valve, × 1, LSJU cat. no. 8697 from Rocky Gulch, 2½ miles SW of Hornbrook, Siskiyou Co., California, photo by Takeo Susuki. Compare flank sculpture to that of figs. 5 and 7 which are also from Rocky Gulch.

Figs. 2, 5. Hypotype, UCLA cat. no. 38761 from UCLA loc. 6140; fig. 2, both valves, × 1, beak view showing anterior profile, beak sculpture and fine ribbing of corcelet and escutcheon which are distinctly different from that of *Yaadia* spp.; fig. 5, right valve, × ¾.

Figs. 3-4. Hypotype, left valve, UCLA cat. no. 38762 from UCLA loc. 6305; fig. 3, × 1, beak view showing sculpture of corcelet and escutcheon; fig. 4, × ¾, flank.

Figs. 6-7. Hypotype, UCLA cat. no. 38760 from UCLA loc. 6140; fig. 6, both valves, × 1, beak view showing sculpture of corcelet and escutcheon; fig. 7, right valve, × ¾, flank with nodes coalesced to form divaricate rib pattern.

ISBN: 0-520-09582-0